Cosmos

Cosmos

RICHARD J. PENDERGAST

New York
Fordham University Press
1973

Printed in the United States of America

To
KARL RAHNER and PIERRE TEILHARD DE CHARDIN
masters and older brothers in Christ

Contents

Preface   ix

1 The System of the World   1
    1. *The hierarchical structure of the universe*
    2. *General system theory*
    3. *Reductionism vs. holism*
    4. *The holistic character of human existence*
    5. *Implicit knowledge of human values*
    6. *The relation of part to whole*
    7. *Holism on the sentient level*
    8. *The motivation of reductionistic scientists*
    9. *Holism on the vegetative level*
    10. *Science and metaphysics*
    11. *Mind and computers*
    12. *The holism of nonliving systems*

2 Evolution   41
    1. *Holism and evolution*
    2. *Neo-Darwinism*
    3. *Teleology and evolution*
    4. *Time*
    5. *The beginning of time*

3 A Model for Space, Time, and Matter   61
    1. *The digraph model*
    2. *Relation of model to relativity theory*
    3. *Local and holistic aspects of development*
    4. *Extension of the digraph model to higher-level
       systems*

4 Personal Knowledge   83
    1. *Piaget's theory of intellectual development*
    2. *Kuhn's theory of the development of science*
    3. *The act of creation according to Koestler*
    4. *Personal knowledge*

5 Symbolizing Activity   103
    1. *The primordial heuristic anticipation*
    2. *The notion of being*

3. *The Trinitarian origin of being*
4. *The theory of symbolizing activity*
5. *Knowledge and decision as symbolizing activity*
6. *Truth*
7. *Knowledge and faith*

6 Creation, Predestination, Spirit   127
1. *Creation*
2. *Predestination and the historicity of God*
3. *Matter and spirit*

7 The Problem of Evil   143
1. *Evil: problem, mystery, and testing*
2. *The ground of the cosmos*
3. *Dualism and the problem of evil*
4. *The traditional doctrine of Original Sin*
5. *Concupiscence and aggression*
6. *Scripture and the problem of evil*
7. *The fall of the cosmos*
8. *The rebellion of modern man*
9. *The existence of Cosmic Powers*
10. *A happy fault?*

8 Hope   183
1. *The future of man*
2. *The shape of the future*
3. *The end of time*
4. *The final word of God*
5. *The works of man*
6. *The hope which is ourselves*

Bibliography   203

Preface

THIS BOOK IS ABOUT THE COSMOS, the ordered totality of all that exists outside of God, and about God Himself insofar as He is the source of the cosmos. Since it is only a few hundred pages long it clearly does not say everything which could be said about the cosmos; all the libraries of the world do not do that. Instead it presents a unified theory of its most general features, discussing details mainly for the sake of the light they cast on the whole.

I have wanted to write such a book ever since I first read Pierre Teilhard de Chardin's *Phenomenon of Man* in 1958. I first began to think that this was really possible when I became familiar with the ideas of Karl Rahner in 1961. Over the intervening years I have learned, somewhat to my surprise, that in this age of confusion a new world-outlook is building quietly and almost invisibly. The ideas of a number of outstanding modern thinkers fit remarkably well into the framework provided by Rahner's notion of being and symbol and Teilhard's insistence (along with Whitehead and others) that dynamic process is a fundamental feature of cosmic reality.

Thomas Kuhn has described periods of groping and confusion in the history of science which he calls "scientific revolutions." Ever since the dissolution of the medieval world-view in the sixteenth century, Western man has been struggling through a period of cultural revolution. I believe that we are now considerably closer to the end than to the beginning of this period. There are signs that the accumulating experience of the race, which has been progressively enriching (even while confusing) our collective subjective consciousness, must one day arc over into a new and more adequate explicit symbolic representation of reality. Unless and until this happens, none of the practical actions taken to resolve the crises of our times can amount to anything more than a bandage applied to a near-lethal wound. Man is the symbol-making animal; without adequate symbols he languishes and perhaps dies. A new vision of reality as a whole is needed in order that men may live together in love and cooperate with one another on the basis of a shared understanding of the world. My hope is that this book will make a contribution to the emergence of such a vision.

Perhaps a description of the contents of the book will be helpful. The first three chapters are "cosmological" in a narrower sense than are the rest. Chapter 1 deals with the hierarchical structure of the material universe, Chapter 2 with its evolution, and Chapter 3 with a quasi-mathematical model for space, time, and matter which helps to concretize the ideas. The hierarchical structure and evolutionary development of the universe raise the question of the nature of the relationship between the various levels of the hierarchy. Hence a recurrent theme of these chapters is an antireductionist polemic and the insistence that holism, teleology, and the recognition of the primary importance of human values and metaphysical concepts are vital for a correct understanding of the world. The acceptance of holism raises the question of the way in which beings of greater ontological worth such as man can emerge from those of less. If causality is to be maintained, as it must, the answer must be given in terms of the creative activity of some cosmic ground which is the source of the novel value which emerges at each evolutionary step.

Chapter 4 begins with a discussion of the work of Piaget and his school. His discovery that human intelligence is hierarchically structured and evolves through stages of increasing structural complexity suggests a close connection between cosmic evolution and individual intellectual development. The ideas of Kuhn, Koestler, and Polanyi on the development of science are considered next. There are analogies between the intellectual development of the individual and that of the scientific community. The work of Polanyi makes it clear that the structures which emerge in the course of both developments are not the whole story and that they depend upon and must be understood in terms of the subjective and personal reality of the human knower.

The insights and problems which emerge in the fourth chapter are understood more deeply in Chapter 5 in terms of Rahner's theory of symbolizing activity. A brief statement of the fundamental concept is: "All beings are by their nature symbolic, because they necessarily 'express' themselves in order to attain their own nature." Thus "each being forms, in its own way, more or less perfectly according to its own degree of being, something distinct from itself and yet one with itself, 'for' its own fulfillment." [1] Symbolizing activity and causality are identified, and *being* is defined as that which is the active or passive term of symbolizing activity. Man's primordial heuristic anticipations, which constitute the horizon within which beings can appear to him, are thus understood as the symbolic terms of his constitutive relationships with God and cosmos.

The theory of symbolizing activity is grounded in a dialectic between our human experience of knowing and desiring and the Christian revelation about the Trinitarian relationships of the three divine persons. This means that the naturalistic level of discussion of the first four chapters is definitely abandoned and that theological considerations play an important role for the remainder of the book. This is unfortunate in some respects since a goodly portion of Chapters 5 through 8, which could have been developed without appealing to revelation, is now "contaminated" for non-Christians. However, from my Christian standpoint the gain in clarity, cohesiveness, and persuasive power obtained by grounding the ideas in revelation more than compensates for this. Non-Christian readers can simply regard the postulate as something which arises from a myth-impregnated form of experience and then make use of it much as I do.

The sixth chapter elucidates certain aspects of the relationship between God and the world and between matter and spirit in terms of the notion of symbolizing activity.

The identification of causality with symbolizing activity leads in Chapter 7 to the realization that the reason for the structural analogy between cosmic evolution and personal intellectual development is the fact that both are symbolizing activities and that evolution is therefore grounded in a personal knower or knowers whose symbolic production is the material universe. The fact that evolutionary development is an autonomous activity of the cosmos itself, together with the existence in it of evil, requires that the Cosmic Powers which ground the universe be distinct from God.

The discussion of the problem of evil is developed further in relation to Scripture and the traditional Christian notion of Original Sin. As the writings of Camus illustrate, this problem is the major obstacle to modern man's understanding of the cosmos, from a practical as well as a theoretical point of view. My solution is a modified dualism based partly on suggestions from the Bible and partly on the philosophical framework developed in the earlier part of the book.

Finally, in the eighth and last chapter, the future of the cosmos and the nature of man's hope is seen in the light of the notions developed in the preceding chapters, particularly that of "heuristic anticipation."

The heart of the book is found in the fifth and seventh chapters. The least essential chapter is the third, which may cause some trouble for those who are not accustomed to thinking in terms of mathematical models (even though there is nothing of special intrinsic difficulty in it). The most unfortunate omission is a detailed discussion of the theory of knowledge and, more particularly, the theory of our knowledge of God. However, it would be hard to go into the topic any further without doubling the size of the book, so I have decided to save what I have to say about it for another work.

A word about references: Practically all the works cited in this book are listed in the Bibliography. When I wish to cite a particular work, I refer to it by giving the capitalized name of its senior author, e.g., "LONERGAN, esp. pp. 205, 6." If more than one work of the same author is listed in the Bibliography, the particular work in question is distinguished from the others in both reference and Bibliography by means of a Roman numcral, e.g., "WHITEHEAD (II), p. 72ff." I hope I will be forgiven some grammatical irregularities which arise when I wish to refer to an author and one of his works simultaneously, e.g., ". . . theory . . . by BRANS and Dicke."

I wish to express my gratitude to St. Peter's College which allowed me a sabbatical leave to finish this book.

NOTE

1. RAHNER (III), pp. 224, 228.

Cosmos

1

The System of the World

1. THE HIERARCHICAL STRUCTURE OF THE UNIVERSE

THE KNOWLEDGE EXPLOSION which began in the sixteenth century has by now greatly enlarged our understanding of the universe. If we were to imagine that some catastrophe were about to destroy all that accumulated knowledge and that we could pass on to the next generation only one brief statement about its content, what would we say? Perhaps the most illuminating brief statement which could be made would be this: The universe is hierarchically organized in a sequence of levels in such a way that the entities on any given level are composed of a finite number of subentities from the next lower level which are united and organized by characteristic interactions. Thus organisms are comprised of cells, cells of molecules, molecules of atoms, atoms of elementary particles. It is important to note that on each level of the hierarchy the reality of an entity includes not only the reality of the subentities which compose it but also that of the interactions between them.

Let us use the term "system" to denote a set of entities united in a coherent pattern by interactions with one another. We can say then that the universe is hierarchically organized in a

1

sequence of systems, in which the systems of each level are composed of subsystems from the next lower level.

As one proceeds up the hierarchy from elementary particles to organisms, there is a growth in complexity. A system which enters into a higher-level system does not lose its characteristic structure and properties. Rather, these are presupposed and used in the formation of the superordinate system. The higher-level system organizes its subsystems in a qualitatively new way which is harder to describe completely and which functions in a way which is harder to predict with precision.

There is more individuality on the higher levels. This is true, first of all, in a quantitative sense. There are far more different structures possible on the level of the organisms than there are on that of the atoms and, at the same time, there are far fewer individuals available to realize those structures. As a result, there are some possible species which have no living representative, and very probably there are no two organisms of even moderate complexity which are completely alike. The higher-level systems seem to be more individual in a qualitative sense also. Each man, and even each dog, has a personality of his own in a sense which cannot be applied to electrons.

The structure of the levels themselves becomes more complex as one goes up the hierarchy. There is little difficulty in distinguishing elementary particles from atoms and atoms from molecules. But complex molecules shade imperceptibly into viruses which themselves shade into complete living organisms. Again, the levels of the elementary particles and the atoms are relatively homogeneous. But the level of the molecules begins the development of branches with the strong differences between inorganic molecules, organic molecules, and crystals. By the time one arrives at the level of the organisms, relationships are very complicated indeed.

We can distinguish three sublevels among organisms: the vegetative, the sentient, and the intellectual-volitional. Plants are the purest representatives of the vegetative level. They are organized for assimilation, growth, and reproduction, but have limited sentience, that is to say, limited power to recognize and respond to the environment. Higher animals other than man are the purest representatives of the sentient level. Like the plants they grow and reproduce, but they also have the power to perceive, respond to, and learn from, their environment. We shall argue later that this capability results from their having another level of organization in addition to that which takes care of growth and reproduction. Finally, we reach the intellectual-volitional level in our own race, its sole representative on this planet. Man is set apart by his intellectual and volitional powers which (we shall argue later) mark the advent of still another level of organization beyond what is needed for mere sentience.

Though we can thus distinguish different sublevels within the group of organisms, these levels overlap considerably. Some plants are more highly developed than some animals, and students of evolution are at pains to determine where the properly human level begins. There are a great many different ways of being an animal or a plant, and these forms of organization merge into one another in complicated ways. Thus, although our brief statement about the hierarchical organization of the universe is indeed an illuminating one, it needs to be qualified. It seems to have about as much validity as the division of human knowledge into physics, chemistry, biology, psychology, sociology, etc.

If we look upward from the level of organisms, still further levels are distinguishable. Animals and men unite through social interaction into societies. These societies interact with one another and with the nonliving environment to form partial ecosystems and finally the all-inclusive geosystem which is this earth. The earth in turn is part of our solar system, which is part of our galaxy, which is part of the local group of galaxies, which is part of the universe.

It is doubtful, however, whether the term "upward" in the preceding paragraph was the correct one to use. Is society indeed "higher" than the individual? This relationship is more subtle than that between a molecule and an atom and will need much more careful discussion. Certainly it would seem strange to call a solar system a "higher" type of system than a human society. The interaction which organizes it is clearly less complex than that which holds society together, and the type of organization is far simpler. The structure of the solar system depends only on the gravitational interaction, which would be the same if the gravitating bodies supported no life at all. This structure contrasts sharply with the organization of an animal which makes use of the properties of the cells and molecules within it. Animal organization is higher than cellular or molecular organization because it organizes the highest properties of those lower levels. The solar system, on the other hand, includes bodies which are alive, but it does not organize their vital activities, only their gravitational ones. It seems clear that either the individual man or the human society is the "highest," most complexly organized, entity which we know, even though each is included within larger entities.

2. GENERAL SYSTEM THEORY

This basic structural pattern which exists throughout the universe is very striking. One feels intuitively that it points to something quite fundamental which has not yet been clearly grasped, but which, it is fairly obvious, has some connection with evolution. Most of the levels of the hierarchical structure have been built up by the evolutionary process. So the question, why does the same structural pattern appear in entities of all sizes, is related to, and perhaps reducible to, the question about the nature of the evolutionary process by which they are built up.

None of the traditional disciplines is capable of dealing with these questions, which cut across all of them. During the past twenty-five years, however, a new field of study, "General System Theory" (GST), has been developing.[1] GST has been prompted, at least in part, by the recognition of the structural similarities we are discussing. It aims to develop concepts which are applicable to all the traditional disciplines and thus to provide them with a unifying framework. Some of its practitioners, no doubt, also hope that it will succeed in uncovering the ultimate reasons for the developing hierarchical structure of the universe, perhaps in terms of some general mathematical theory of interaction and growth.

But at present GST is a rather loose constellation of concepts, methods, attitudes, interests, problems, and some results. The people interested in it sometimes have very different fundamental orientations and very different ideas of what it is. This is partly the result of the fact that GST is closer to philosophy than to the traditional disciplines, and thus one's views of it are inevitably influenced by philosophical convictions. Nevertheless, I think that it is true to say that those active in GST would like to keep it nonphilosophical insofar as it is possible. One of the great advantages is the very fact that it provides a meeting ground for people whose philosophical positions are quite disparate.

The central concept of GST is that of "system." As the term "solar system" reminds us, this notion is as old as the science of physics. The physicist is accustomed to dividing the set of entities of interest into two subsets, the "system" and its "environment" or "surroundings." This division can be done quite arbitrarily, but in general the system will include the entities whose behavior one wants to study and most of the important interactions. Interactions across the "boundary" between the system and its environment will often be represented only in an average way in terms of "boundary conditions." Thus the system tends to correspond to natural unities, or natural systems, similar to those in the hierarchy we have discussed.

The concept of system has been enriched by modern engineering. Unlike physical scientists, engineers not only analyze systems, they also synthesize them in accord with some human purpose.

A machine is the embodiment in matter of a "rule of rightness" determined by that purpose.[2] Two physico-chemical systems might be quite disparate from the viewpoint of physics and chemistry and yet be identified as essentially the same machine in a court of patent law. Consequently, the study of engineering systems necessarily involves the concept of purpose.

However, a machine as such exists only in the minds of the men who know it. The physical object which is a machine for us is in itself a mere physico-chemical system without any intrinsic purpose. Its purpose and meaning as a machine are projected into it by our minds and are not part of its physical reality. That physical reality can be explained completely (at least "in principle") by the laws of physics and chemistry. I am making here the classical distinction between what is known and the manner in which we know it.[3] Though the object we know as a machine is totally explicable in terms of physics and chemistry, the engineer knows this object in relation to human needs and purposes, a relationship which cannot be explained in terms of physics and chemistry.

The role of purpose in the constitution of machines has become more strikingly evident since the Second World War and the advent of self-regulating machines. These machines are equipped with devices ("sensors") which can detect the state of the environment and the internal state of the machine. If the state of the machine is not what is called for in relation to the environment, an "error signal" is generated and fed back to the input side of the machine, causing it to move toward the desired condition. The study of such self-regulating machines has been called "cybernetics." Clearly enough, the cyberneticist needs the concepts of purpose and goal.

A new stage in the evolution of machines has been reached with the modern "computer," which not only computes but also "recognizes" patterns, "makes decisions," and, in due time, will probably become capable of carrying out any operation, even the most seemingly intellectual, which a human being can specify completely in terms of a finite number of operations. Cybernetics is now becoming the study of the "behavior" of automata and of "artificial intelligence."

3. REDUCTIONISM *vs.* HOLISM

These developments in modern technology have supplied valuable analogies for neurophysiologists and biologists in general, and have contributed to a tendency to think of organisms in terms of the same concepts as machines. The interest of psychologists and social scientists in the system concept has also been aroused, and notions like "feedback," "stability," and "information" are now commonplace among them. This is a good thing in itself, but unfortunately the drive for unity sometimes short-circuits into one of the major intellectual vices of our day, "reductionism."

By reductionism I mean the philosophical position which claims that the more complex systems in the natural hierarchy, including man himself, can be reduced to nonliving ones governed by the laws of physics and chemistry. In our day this reduction is usually accomplished with the aid of the intermediate stage of the cybernetic machine. It is clear, as Polanyi has pointed out, that we understand biological systems as wholes before we analyze them. Just as the technologist is not interested in every physico-chemical system but only in those which are related in a suitable way to human needs, so the biologist is not interested in every natural system but only those which he intuitively recognizes as interesting because they have "life." Biology begins with classification, an art dependent upon connoisseurship of a high order which intuitively grasps the object as a whole and sees the parts in terms of the whole. The art of the taxonomist is but the development of the intuitive interest of all men in living things. No one who has spent a half-hour at the zoo attentively contemplating a caged lion can come away without a somewhat reverential impression that here is a highly individual, interesting, and beautiful, albeit frightening, entity. No doubt this intuitive way of understanding animals as wholes was a necessary evolutionary development without which our species could not have survived. Until the advent of modern science, this intuitive reaction was taken as clear proof that animals exist as irreducible wholes which cannot be completely explained in terms of their parts. Reductionism, however, denies this and asserts that just as we must distinguish between the mode of existence which a machine has in itself and our way of understanding it, so we must distinguish between the mode of existence of the living system and our way of understanding it. Our study of living systems is begun with a global recognition of the animal as a whole, but scientific analysis enables us to separate the elements of our understanding which relate to the thing in itself from those which result from our method of knowing it.

Many system theorists are reductionists, but another viewpoint is possible within GST and is in fact held by many. I shall call it the "holistic" or "organismic" viewpoint. Ludwig von Bertalanffy, one of the founders of GST, argues for this viewpoint when he writes that "the objections against a 'machine theory' of life remain valid" and that a fully developed

GST "would indeed incorporate the 'organismic' world view of our time, with its emphasis on problems of wholeness, organization, directiveness, etc., in a similar way as when previous philosophies have presented a mathematical world view (philosophies *more geometrico*), a physicalistic one (the mechanistic philosophy based upon classical physics), etc., corresponding to scientific development."[4] A full-blown holistic view would argue that in fact some higher-level systems cannot be reduced to lower-level ones and that the former surpass the latter in being and value.

This is not true, of course, of all systems. There are systems which are mere conglomerates, exhibiting "unorganized complexity": A collection of molecules in the gaseous state, for example. We may think of such a collection as a single object, but in itself it has little or no unity. The intellectual model by which we understand it is indeed a unified conception, but its unity is the product of our minds and does not touch the system itself.

However, I shall argue that all the systems in the following sequence are irreducible units which cannot be completely explained in terms of their parts: man, brute animal, plant, cell, complex molecule, atom, elementary particle. The general structure of the argument is as follows: I show first that the data of personal experience rule out reductionism as an explanation of man. This I consider certain. Next, I argue that when we interpret the data we have about sentient animals within the framework of our experience of our own animality, it appears highly unlikely that reductionism can be correct for them. The same is true to a lesser extent for plants. Finally, I suggest that coherence requires that we understand all natural systems in terms of the framework developed for living things. The remarkable fact of the hierarchical organization of matter points to something quite fundamental in the nature of the universe. A basic principle is at work which expresses itself in an analogous way on all levels, and so what we know about the higher levels is probably true of the lower ones as well.

The pattern of my argument is the opposite of the reductionist's. He begins by making the mistake of supposing that his knowledge of lower-level systems is not only clearer but also more certain and more valuable than his knowledge of higher-level systems. Consequently, he has no hesitation in putting aside basic, but unclear, intuitions about complex entities like men and animals in favor of hypotheses derived in part from his notions about atoms. The rise of science began in the sixteenth century with spectacular progress in physics and astronomy, which was soon followed by further breakthroughs in chemistry. As a result the "hard sciences," particularly physics, were taken as the ideal model of what a science should be. Although there were always respectable scientists who opposed the idea that biological and social phenomena could be totally explained in terms of the laws of physics, there was nevertheless a tendency to model the conceptual structures of other disciplines upon those of physics. During the past one hundred years, however, biology, psychology and, to a lesser extent, the social sciences, have developed greatly, and engineering has become a highly theoretical enterprise. All these subjects have evolved important concepts which are relatively independent of physics, and so the realization has grown that whether or not everything can be reduced to physics "in principle," workers in other disciplines must as a matter of fact create and use basic ideas which have roots in their own field of study and are not derived from physics. Furthermore, the suspicion is growing that physics may not have privileged access to reality and that the models of physics may not reflect it any more clearly than the models of other disciplines.[5] For my part I am sure that if mankind does succeed eventually in constructing a comprehensive model of the physical universe, some of its fundamental features will depend upon evidence not available to the physicist, which will be obtained by study of living organisms and societies.

4. THE HOLISTIC CHARACTER OF HUMAN EXISTENCE

As a prelude to the discussion of holism let us try to understand the nature of the machine more deeply. The modern machine, like a man, is an example of "organized complexity." It has many parts exhibiting a high degree of organization, but the organization is still an "accidental" one which does not penetrate or alter the inner nature of the parts. Rather, the parts remain what they were before they were assembled into the machine and act according to the same physical laws which they obeyed before they became parts. What has happened to them is that "boundary conditions" or "constraints" have been imposed on them which restrict their action and reduce the number of degrees of freedom of the whole set of parts below the number it would have if they were completely independent.[6] In this way the total assembly is restricted to states which are desirable from the human viewpoint of the machine's designer. However, the action of the totality is not qualitatively superior to the action of the parts, even though it is controlled and constrained into patterns which are highly improbable from the viewpoint of inanimate matter alone.

A savage from the interior of New Guinea who might unexpectedly come upon a very complex modern machine might well imagine it to be endowed with a "soul," some intangible principle over and above its inanimate parts, which would enable it to perform in such a flexible and responsive way. But he would be wrong in his supposition. Adequate knowledge of the laws governing inanimate matter is sufficient to enable one to deduce the behavior of the total ensemble. In the case of a machine, there is absolutely no need for the hypothesis of a soul. This is a particular example of a general principle: If one can deduce all the features of the whole from the behavior of the parts as independent entities, then there is no need to suppose that the whole is anything more than a "generalized machine." The deduction establishes a reduction.

The living system which is man also exhibits "organized complexity." But the organization here differs qualitatively from that of a machine. We have direct experience of our own existence in and for ourselves. We are aware of ourselves performing various actions. We see, hear, touch, desire, choose, know, love, etc. All these various performances of ours are extensions and modalities of a single basic performance, that of be-ing. At various times I am running, laughing, weeping, loving, knowing, etc., but always under and in each of these I am. It is in relation to this basic performance of being that I understand myself. I am he who exists in the way with which I am so intimately, though confusedly, cognizant. This way of existing is "substantial" and also "personal." I exist in and for myself, and I have a certain genuine, though limited, independence. My being is my own performance, and it cannot be attributed to anything outside of me or to any part of me, except in relation to, and dependence upon, the integral me. It is precisely because of my self-performance of being and its extensions in those other performances of knowing, loving, etc., that I have a unique and irreplaceable value: no one else can ever do what I do, viz., be myself and as such know, love, etc., in my own incommunicable way.

In all our performances we are directly aware of the object of the action and subjectively aware of ourselves performing them. Thus consciousness has both a subjective and an objective pole. There are two terms of the relational reality we call knowing: objective knowledge and subjective awareness.[7] Objective knowledge is clear and distinct, subjective awareness is not. The personal subject as such is never the object of his own knowledge; one knows oneself as the source of the act, as the one who is attending to something other than oneself. This is true even when the subject reflects upon himself. Reflection is an act in which the subject forms objective knowledge of himself. He does not cease to be aware of himself subjectively, but he creates an objective symbol in which he also knows himself objectively. This point is worth noting. Piaget and others have spoken at times as though the human subject could exist without being aware of himself as distinct from his world. To maintain this in an unqualified way is to confuse subjective awareness and objective knowledge. If he knows anything at all, the subject is subjectively aware of himself as knower. But he may not know himself objectively as distinct from the world. The research of Piaget (discussed below) has shown that in the beginning the objective knowledge of the child mingles the self and his world in a single, confused whole. Even the objectively known self is "other" with respect to the subject, and the self-image is not the same as subjective awareness of self.

In later chapters I shall discuss at length the nature of

human knowledge. I shall argue that, in the act of knowing, an objective symbol proceeds from the subject who has been actualized by God and the world. The symbol produced is objective knowledge; the subject's awareness of self in contact with the real can be referred to as subjective knowledge. It is also implicit knowledge; knowledge implicit in the self-awareness of the subject. This type of knowledge, a knowledge identical with or contained in self-awareness, seems to approximate what Michael Polanyi intends when he speaks of "tacit knowing" or "subsidiary awareness," and I shall use these terms frequently. Of course, my use of the terms is not simply identical with Polanyi's, since my frame of reference is different, though not, I believe, incompatible.

The unitary and substantial character of personal existence is known to each person with the utmost certitude. But this knowledge is in the first instance subjective and implicit. It is so much a part of my conscious experience that I assume and build upon it without any necessity of adverting to it explicitly. If I do advert to it, I can express it, but only in terms of experience. One can speak to others about it only by referring them to their own experience and their own implicit, lived, knowledge. Consequently a person who has been conditioned to regard only what is "scientific" as important, and only what is clear and objective as scientific, will be tempted to overlook his own reality as a person, even though he implicitly knows it in everything that he does.

Once a person explicitates his subjective awareness of his own being, it is immediately clear that reductionism cannot be correct. It is the integral I who live, know, and love. These operations and the subject who performs them cannot be reduced to anything simpler. Furthermore the personal subject and his human acts are better known and of higher value than any of the substructures which common sense or science may discover in man. The reductionist is guilty of the fallacy of "misplaced concreteness." [8] Either intellectually or by physical means he excises a part or aspect from the concrete whole and attributes primary existence and meaning to it, ignoring the certain, albeit implicit, knowledge of his own primary reality which he exhibits in the very act of knowing.

5. IMPLICIT KNOWLEDGE OF HUMAN VALUES

I subscribe in large part to the epistemological views of Michael Polanyi. Polanyi points out that all our acts of knowledge, including scientific knowledge, are skillful performances which suppose as implicitly known much more than is expressed.[9] One can draw an analogy from athletics. In order to perform skillfully an athlete must coordinate all the parts of his body with precision and must therefore be aware implicitly of their position and how it changes. Yet he could not possibly convey this information to a novice, nor would it be desirable for him to do so. Explicit attention must be focused on the performance as a whole, and focal awareness of details would make this impossible. It is true that the learning process may sometimes be facilitated by attending to details in an explicit way during practice. But this objective awareness must be reintegrated into subjective, implicit self-awareness when peak performance is desired.

When the scientist performs, he knows implicitly much more than he can tell. In particular he knows and is motivated by the being, truth, beauty, and goodness of the physical universe. Being, truth, beauty, and goodness are human values, known intellectually by human acts which cannot be reduced to anything below the human level. Reductionism is therefore incompatible with the very foundations of scientific activity within the personality of the scientist himself. If reductionism were true, that which the scientist wants when he practices science would be meaningless.

Kuhn has shown that the creation of a new scientific viewpoint is a nonscientific act, in the sense that the new conceptual system is not dictated by the data available and often contradicts the existing canons of science.[10] The scientific innovator remakes science, and this remaking is not sanctioned by any explicitly known criterion of what is scientific. The obvious and frequently used illustrations are relativity and quantum theory.

This manner of speaking may perhaps be justly described as somewhat strained. Surely the act for which a scientist is above all admired must be "scientific." But if it is, then science is incurably intuitive and unformalizable and akin therefore to the philosophical and even to the metaphysical and religious. One must assert that scientific activity is a noble human enterprise which involves the entire person with his values and moral qualities.

Most reductionists never advert to this. Many of them are good men who in practice are as tenacious of human values as are other men. But the cultural and educational system of our pathologically specialized and splintered age has put them into a strange situation. A genuinely intellectual man should aim at the acquisition of an explicitly known conceptual system which is as all-inclusive and as unified as possible. Obviously this conceptual system should be organized in such a way that the most important aspects of reality are central to it and subsidiary aspects are understood in relation to what is central. But the religious and moral confusion of our times has obscured the concepts which express explicitly the implicit knowledge which all men have of the central human values. At the same time there has been a vast growth in the conceptual apparatus available for handling subsidiary aspects of reality. As a result, the young student finds that he can acquire rapidly and successfully a conceptual apparatus which deals with some limited and subsidiary aspect of reality, while the concepts needed for dealing with central human and religious values are either nonexistent in his milieu or are far more difficult to acquire. The result in many cases is an unthematic (i.e., lived and implicit) decision to be intellectual only in regard to a limited area of reality and not to reflect in a serious or sustained way on the central human and religious values and structures of the universe. When large numbers of naturally gifted men make such a decision, a bizarre situation is created. The specialized disciplines whose subject matter is most remote from man flourish, while those which are closer to him are crippled by the eclipse of religious and philosophical meaning. The result is that the more uncontaminated a concept is by relevance to what is central in reality the better the chance it has of becoming normative. Central human experiences are forced into a Procrustean conceptual framework and are never fully understood explicitly.

A morally good man will, of course, have implicit, lived knowledge of the central human realities but if his explicit conceptual apparatus is inadequate for dealing with them, definite strains are introduced into his moral life. There is danger that his human personality will be stultified by a loss of interest in human values or that he will come to consider the life of the mind as of small value. This second danger is apparently now being realized among some young people. The present era has been compared to that of the Reformation. During the sixteenth century there was widespread loss of faith in various beliefs which had formed part of the cultural fabric. Today there seems to be a loss of faith in the relevance and worth of intellectual discipline and of science in particular.

This reaction I regard as far more logical (though less valid, in some respects at least) than the attitude of the reductionistic scientist or scholar. He knows implicitly the various human ideals which motivate him, but the conceptual system he explicitly holds reflects these ideals very inadequately. If he were really doing what his conceptual system says he is, then both he and his science would have only trivial importance. But fortunately he is not very logical (except within the confines of his discipline), so he continues the important human enterprise for which he can give no persuasive explicit reasons.

The young, on the other hand, though much less capable of logic within any particular discipline, are not afraid to apply it to larger issues and to conclude that any intellectual enterprise which reduces human experience to the lowest level is not very important for human beings. Love, truth, beauty, and goodness are what we want and need, so why value an enterprise which dissolves them into loveless blind facticity?

6. THE RELATION OF PART TO WHOLE

Granted that the data of our personal experience are incompatible with reductionism: how are we to understand the relationship between ourselves and the various structures of our bodies and minds? How are the parts related to the whole? There are other inadequate answers besides reductionism. One of them was given by Descartes. His concept of man has been

level systems do not describe their behavior so accurately after these systems become dependent upon a superordinate entity into which they have been integrated. But for present purposes it makes very little difference whether entry into a higher-order system specifies the indeterminacy of a lower-level ideal law while leaving the ideal law intact or whether it modifies the ideal law itself. Indeed, the two views may be nothing more than different ways of expressing the same thing.

All this is related to the perplexing question of how to interpret the mathematical formalism of quantum theory. "Despite its enormous practical success, quantum theory is so contrary to intuition that, even after 45 years, the experts themselves still do not agree what to make of it. . . ." [13] The rules for making calculations are well known and give consistent results, but nobody with an iota of philosophical acumen can use them without being tantalized almost beyond endurance by the question of what it all means ontologically. Responses tend to fall into two different classes, much as they do for the problem of God. There are the "tough-minded" agnostics who say "We don't know and we don't need to know," and the others who come up with a set of wildly varying conjectures. It is interesting that the notion of consciousness is central to some of the most important ones. Heisenberg, who started it all, espouses a personalized version of the "Copenhagen interpretation" which seems to be a cross between Aristotle and German idealism. Wigner, one of a few whose work has most powerfully affected the contemporary form of quantum theory, holds that the interaction of a physical system with a conscious observer changes its state in a way in which other interactions do not.

Everett, Wheeler, and Graham (EWG) offer another interpretation which involves a realistic explanation of the formalism (as against Heisenberg) and the rejection of a special role for consciousness (as against Wigner). The result is the severely counter-intuitional hypothesis of a multi-branched cosmos in which every interaction which, according to the formalism, can yield more than one result produces a split of the cosmos into a multiplicity of branches, in each of which one of the possible results of the interaction is realized. Thus, for EWG there are at this moment billions of different Pendergasts, each conscious of himself in the state produced by the chain of interactions proper to his own branch of cosmic history.

Still another interpretation is that of Bohm who postulates the existence of "hidden variables" which, when understood, will reveal that the indeterminacy of present-day theory is but the reflection of our ignorance of the details of deterministic physical processes underlying the present quantum level.

For my part, I believe that there are probably many as yet unknown "hidden variables" in nature. I doubt that I differ in this from the majority of physicists. Even a cursory acquaintance with the history of physics discourages any notion that our present theories are final or that nature does not still have many surprises in store for us. But belief in the existence of as yet unknown physical structures is different from the assertion that the fundamental processes of nature are deterministic. It is indeed true, as Einstein insisted, that blind chance cannot be the ultimate determinant in nature, but this does not imply determinism. Rather, there is an irreducible creative element in nature. The emergence of each new moment from the past is a creative decision which cannot be predicted even by an omnicompetent calculator. One way of stating the main argument of this book would be to say that at the root of cosmic process is consciousness. The cosmos in its development through time does indeed encounter branching points as EWG assert, but these are moments of decision rather than of schizophrenic splitting. I find it implausible to say that human consciousness can change the state of a physical system under observation, but there are other consciousnesses in the cosmos. Physical process has two aspects: one is mechanistic and can be expressed in terms of rigid laws, the other is creative. This creativity is the ultimate source of all process and of all the mechanisms it uses for its own purposes.

The indeterminacy of present-day quantum theory may well be resolved on a much more superficial level than that of the ultimate creativity at the root of cosmic process. But it is at least interesting to observe that the most fundamental theory of modern science gives little encouragement to reductionistic thinking.

Holism does not deny then that living systems contain a

great deal of "living machinery" which operates, in the first approximation, according to physico-chemical laws.[14] But it goes further in asserting that they also display "organismic self-regulation" by which the whole determines the activities of the parts in a nonmechanistic way.[15] It is by organismic self-regulation that the total system builds up and integrates the various mechanisms which it needs in order to function for its own irreducible ends.

The most important example of organismic self-regulation is the act by which a man configures his field of consciousness.[16] It is evident that our acts of perception, desire, knowledge, choice, and love depend upon many mechanisms, known, suspected, and completely unknown, which operate in the brain and central nervous system. And yet the total conscious *Gestalt* is not a simple additive sum of the contributions of all these mechanisms. It is an integral whole which permeates and determines the final value of each of the elements within it. There is an obvious analogy here with a painting or a piece of music. Certainly the various elements which enter the work of art have some independence. One can say certain things about the tonal quality of various instruments or the ordinary effect of various patterns of notes. Nevertheless, the final effect of these elements is strongly determined by the overall pattern of the musical composition into which they enter. Similarly, the field of consciousness of the human person cannot be reduced to the elements which comprise it. The whole is more than the sum of its parts, and though the possibilities open to consciousness at any moment are no doubt limited by the mechanisms upon which it depends, nevertheless consciousness must be understood as a whole if it is to be understood at all. The act by which it is configured cannot be anything other than the organismic self-regulation of the whole personal being which is man.

7. HOLISM ON THE SENTIENT LEVEL

It is certain then that reductionism cannot be a correct explanation of the complex entity which is man. Within the framework set by our understanding of man, we inquire into the nature of sentient nonhuman animals. The higher animals exhibit behavior which naïve human intuition judges to be fundamentally similar to our own. Animals do not speak, do not reason abstractly, do not compose works of art, do not seem to have the same quality of self-awareness as men have; yet they do perceive, desire, decide, and exhibit many of the same emotions. It is conceivable, of course, that in animals these actions are radically different from what they are in our own case, but the fact that we use the same terms to describe them demonstrates the widespread belief that their acts are fundamentally similar to ours. If they are similar, we can understand animal activity and organization in terms of our own. Our experience of perception, sensible desire, joy, satisfaction, fear, shows clearly that these acts are, in us, irreducible wholes, mere elements within our total field of awareness, which is dominated and organized by intellectual factors; yet they have a unity and organization of their own which is not lost by assumption into the totality of human consciousness. Each of these sentient acts depends, no doubt, upon many physiological mechanisms. Nevertheless the contributions of the mechanisms are united in an integral *Gestalt* by organismic self-regulation and thus raised to a level which is properly *sentient*. The visual field, even though extended, is an organized whole in which colors, shapes, sizes, and motions are determined by the entire configuration. The linguistic expression "I see" expresses the unity of visual perception, a unity which is grasped in the very act of perception as stemming from an integral subject.

There is danger, no doubt, in trying to abstract sensible experience from our total conscious experience. Our integral human consciousness is primary, and we can easily distort the nature of sensory and motor acts in trying to separate them from the whole into which they have been assimilated. Nevertheless it does seem possible to assert with confidence that there is a level of organization in man below the personal level and that it is on this level that perception and motor organization occur.

This conclusion based on the data of introspection can be illuminated by the data of modern science. A great mass of observations and experiments by ethologists, neurophysiologists,

and psychologists indicates that there is hierarchical organization in the nervous systems and behavior patterns of animals
and that a similar organization is present within man.[17] The
work of the developmental psychologist Piaget and his followers
is especially pertinent. Piaget has studied the development of
intelligence in children. He finds that it occurs in three main
periods, the sensorimotor period (years 0–2), the period of concrete operations (years 2–11), and the period of formal operations (years 11–15). Each of these periods in turn exhibits
various stages. For Piaget, acts of intelligence are indeed actions.
During the sensorimotor period these actions are directly related
to external objects. They are either fully explicit manipulations
or partially schematized imitations. Later, the actions become
progressively internalized until finally they take place mainly in
the brain. The level of organization of the sensorimotor period
is analogous to the highest level attained by nonhuman animals.
It is during this period that we acquire the "sensorimotor concepts" or "schemas" of object, space, time, means–end relationship, and causality which we apparently share with nonhuman
animals.

Mr. and Mrs. Kellogg of Bloomington, Indiana, raised a
young chimpanzee, Gua, together with their own son, Donald,
for more than a year.[18] Although the child showed some
superiority quite early, the development of the two animals
was roughly parallel during the sensorimotor period. The child
began to display marked superiority only with the development
of the symbolic function, which includes speech, toward the
end of the sensorimotor period. It seems that in his ontogenesis
the infant must repeat the development of a type of organization invented by his nonhuman ancestors and continued today
by his nonhuman relatives.

In spite of indications and suasive arguments from science,
the assertion that the analogy between the sensorimotor structure and activities of men and animals implies that animals are
irreducible wholes goes beyond what can be proved by a purely
"objective" thinker. The purely objective point of view does
not enable us to get inside the animal in order to experience the
world from his point of view. And unless we can do that in
some way, the evidence will remain ambiguous. But this
problem about the nature and existence of animals is not the

only one which the objective view cannot decide. Objective evidence alone cannot even determine whether there is such a thing as real causality or whether anything at all exists outside our own mind. As Hume and Kant noticed long ago, invariant sequences of events do not establish causality, and phenomena do not enable us to know things in themselves. The purely objective scientist would be a mere observer and correlator of meaningless sequences of phenomena. He would in fact be a well-programmed computer which could manipulate empty symbols according to certain laws of operation but which could never really know anything. As Polanyi has shown, scientific knowledge, and indeed all human knowing, is rooted in non-objective, implicit, and subjective knowledge. The precise and objective grows out of the richly unclear, unformalizable, and subjective. It is rather amusing to see persons who would insist that there is a real external world, while claiming that the non-objective has no place in science.

A message has meaning only within a structure of possible intelligibilities or "message space." [19] The messages which come to us through scientific observation and experiment have meaning only within the message space or "horizon" of being, truth, goodness, and beauty which is implicitly known to every human being as soon as he interacts with the world. We can empathize ourselves into rocks to see if they exist, and into animals to see if they experience the world in a unitary way, because they and everything which exists are already contained, in a certain sense, within our own being. We have a correct heuristic anticipation of all possible beings in our awareness of ourselves as persons. Within this framework, or heuristic anticipation, the meaning of the abundant evidence available on the organization of nonhuman animals becomes clear. They, like ourselves, are irreducible unities which have their own mode of being and action which cannot be completely explained in terms of anything less highly organized.[20]

8. THE MOTIVATION OF REDUCTIONISTIC SCIENTISTS

This conclusion is only a confirmation of the intuitive judgment which men have always made about animals. It is this implicit judgment which lends its special interest to ethology, an interest very well expressed by K. Z. Lorenz: "I confidently assert that no man, even if he were endowed with a super-human patience, could physically bring himself to stare at fishes, birds or mammals, as persistently as is necessary in order to take stock of the behaviour patterns of a species, unless his eyes were bound to the object of his observation in that spellbound gaze which is not motivated by any conscious effort to gain knowledge, but by that mysterious charm that the beauty of living creatures works on some of us!" [21]

Yet the same Lorenz has spoken as though he believes that not only animals, but men as well, are automata! Here we have a living embodiment of some of the contradictions of the twentieth-century mind. Reading his books, one becomes aware that he loves animals. His whole attitude toward them is quite different from that of an engineer toward a machine. The engineer is delighted with a clever solution to a technical problem; in a sense he loves the machine which embodies it, but he loves the machine only inasmuch as it concretizes an abstract idea. He agrees with the courts of patent law that the individual machine and the particular pieces of iron or plastic of which it is made are nothing; the concept is everything. Lorenz, on the other hand, knows and delights in this particular gray goose, this particular pair of cichlid fish. He knows implicitly that here are unique individual entities which exist as irreducible wholes. His interest in these individuals activates him to seek to find the various mechanisms which enable them to behave as they do. Here the abstract idea acquires value because of its usefulness to the individual entity, just the opposite of what is true in the case of the machine.

Lorenz apparently is not aware that his personal attitudes contradict his theoretical opinion that animals are automata. As far as his scientific work is concerned, this matters not a whit. The concepts he creates are quite appropriately mechanistic because they are the means of understanding mechanisms. Meanwhile his personal appreciation of the unique value of individual animals continues to motivate him to pursue his researches.

Niko Tinbergen, a student and colleague of Lorenz, offers further insight into the motives of reductionistic biologists. In a classic summary of the position of ethology published in 1951, Tinbergen urged that ethologists concentrate their efforts on a "causal" understanding of animal behavior and leave to others the study of its directiveness and of subjective phenomena.[22] He admits that in the case of man subjective consciousness is an undoubted reality, but says that "it is idle either to claim or to deny"[23] its existence in animals. He also seems to understand the undoubted "directedness" of animal behavior in terms of a nonteleological neo-Darwinian theory of evolution which dissolves the genuinely teleological character of animal behavior into the directedness of a machine.

The positive motives behind Tinbergen's attitude are admirable. For far too long people imagined that their understanding of behavior was adequate once they had intuited by empathy its conscious motives. Twentieth-century psychology and ethology have shown that this is far from true. Even those who believe in the irreducible importance of conscious and holistic motivation must admit that it cannot be properly understood unless one has some grasp of the mechanisms which energize it and are integrated by it. I might remark, however, that this can be known to some extent from the data of introspection alone and that Aristotle and the Scholastics of the thirteenth century understood it far better than the philosophers of the nineteenth century, even if not so well as we do.

Einstein once remarked that to understand science it is better to observe what scientists do than what they say. Tinbergen was probably right when he judged that, in the concrete situation of 1951, and perhaps of today also, it would be more scientifically fruitful for ethologists to concentrate on elaborating the mechanisms of behavior than to worry about conscious and holistic motivation. Perhaps we will understand the irreducible emergent properties of the whole soonest by concentrating for a while on the mechanisms. The hypothesis that some particular phenomenon can be explained in terms of the properties and interactions of the parts of the system has been a very fruitful one, and even if it is not completely correct in a given instance, it is usually at least partially right and brings about the elucidation both of what can be explained in terms of mechanisms and of what cannot. However, to absolutize this hypothesis by elevating it to the status of a universal truth is a different matter. In order to do this one must ignore the most important data we have about the nature of reality and must force everything into the Procrustean bed of a conceptual framework designed to handle phenomena which are peripheral in comparison to what is ignored.

9. HOLISM ON THE VEGETATIVE LEVEL

Men are intelligent and sentient, animals are sentient. In addition both share with the plants activities which we may call vegetative, which include assimilation, growth, and reproduction. Do these basic activities of living entities show that the systems performing them are irreducible wholes which cannot be fully understood in terms of their parts? The evidence here is not as compelling as in the case of man and the higher animals. But when it is considered within the framework set by our understanding of the latter, I believe that the answer must be yes.

A word of clarification is needed here. The term "irreducible whole" must be used in quite different senses in referring to man, higher animals, and plants. The organization of a many-celled plant is clearly not so tight as that of a man or even of a chimpanzee. The essential unit for vegetative activity is the single cell, and until we come to the level of sentience, many-celled entities do not add anything as revolutionary to the activities of their cells as what goes on within the cell itself. Later I shall argue at length that there is a principle of unity within the cosmos which determines the activities of all its parts. This determining influence extends to all levels which express the nature of the whole in varying degrees. Consequently, the relationship between different levels is complex, and the unity of each is different in kind. The fact that we have not yet discussed the nature of the whole means that factors vital for a proper understanding of life are still missing, and that what is said about its nature now must be regarded as a first approximation, to be improved on later. At present, however, it can still be validly asserted that living systems and their activities cannot be totally explained in terms of nonliving ones.

The evidence which founds this judgment is vast and multifaceted. It has been ably set forth by a number of different authors.[24] I shall not review it in any detail both because of limitations of time and space and because a fully satisfactory performance of the task requires a vast knowledge of facts and a delicate appreciation of their precise value, which I lack. In general, it can be said that every important vital activity requires a precisely coordinated cooperation of sizable portions of the living system. No one has yet provided a complete explanation of this in terms of mechanism. The cooperative activity seems to well up from within in every part of the organism and, *prima facie* at least, the only thing which can account for this is the determining influence of the whole in each of its parts, an influence which, following Harris, we have called organismic self-regulation.

The argument is partly a negative one: no one has yet given a complete explanation of vital processes in terms of mechanism. Negative arguments are, of course, subject to rebuttal by future developments. In the abstract, one can believe or hope that mechanistic explanations will be developed and can see present scientific advances, such as those in contemporary molecular biology, as earnests of complete success in the future. But the argument is not merely negative. We have seen that man and brute animals are living unities. It seems absurd to suppose that their vegetative activities are not included in that unity and do not share its holistic character. But if basic vital activities are holistic in the animals, it seems that they must be so in the plants as well. There may be great differences, but all living things share the same basic structure, from which flows the activity we call living. The universe is a coherent system, and holism is a fundamental structural principle which applies to all its levels.

The framework within which one considers the evidence is decisive. Scientific study has vastly deepened and broadened the available evidence about the nature of life, but, as far as the fundamental question we are considering is concerned, it has not changed its character. Even the external form of plants and their visible growth and reproduction give rise to a naïve judgment of their irreducible unity. If one maintains the essential horizon within which this naïve judgment is made, further evidence supplied by science does not overturn it, but modifies, elucidates, and confirms it. And, of course, the main burden of this chapter has been that the basic orientation toward human values which the naïve judgment presupposes must be maintained under pain of depersonalization and essential error about the nature of reality.

10. SCIENCE AND METAPHYSICS

Here I shall digress to discuss the attempt by Harris in his excellent book "to found metaphysics in science." Philosophy in general, and metaphysics in particular, is a difficult and risky enterprise, and I heartily agree that it is extremely desirable, almost essential, to tie it in with all the intellectual disciplines and all areas of human experience. Otherwise there is grave danger that philosophy will evaporate into empty concept-grinding which has no genuine contact with the real. Science in particular offers a well-grounded and widely accepted body of facts and theories which can shed much light on philosophical questions. One of the strongest confirmations of the correctness of a philosophical position would be its success in integrating the findings of science into its more universal viewpoint.

I subscribe, however, to the position of "transcendental neo-scholasticism" which maintains that metaphysics is the science which makes explicit the universal implicits which are known to all men in a subjective, unformalized way.[25] Metaphysics does not establish truths which would otherwise be unknown; rather it makes them known in a different way. From this point of view, the metaphysician does not absolutely need anything more than ordinary experience since the universal implicits of human existence are already operative there, though one or another of them may perhaps come to the fore with special poignancy in the specialized experience of a particular historical era or of some particular human enterprise, such as physical science, medical practice, politics, etc.

The great scientist, and, for that matter, the great artist, or the great politician, has superb understanding of some aspects of reality. His behavior is based on this understanding (which, it should be noted, may be only implicitly known to him) and can therefore be a fruitful source of insights for the metaphysician who has the background needed to understand what he is doing. The explicit scientific judgments which he makes are also the expressions of his implicit understanding of reality and can therefore be of great help to the philosopher. However, there are difficulties which make it unwise for the philosopher to rely too much on the apparent metaphysical implications of the scientific judgments of individuals or even of the whole scientific community. These judgments are not really metaphysical and are somewhat incommensurate with metaphysics. They arise from insight into a relatively limited sector of reality, and their formulation is not controlled by a great concern about the way in which they fit in with other data from outside that sector. Science presents the truth in terms of models. This procedure enables science to be objective and clear but at the same time it truncates reality, for no model is simply identical with the reality it represents. It is an approximation accurate up to a point, but there are always aspects or factors of the real situation which are left out of consideration. Perhaps the most important such omission is the fact that things have, as Teilhard said, an "inside" as well as an "outside." No "objective" science has yet managed to take account of human consciousness and the all-important human values of being, truth, beauty, and goodness which are given only in consciousness.

In fact, the question of reductionism versus holism, to which Harris primarily addresses himself, is unanswerable on the plane of science. No one knows enough to deduce either the actual behavior of organisms or its contradictory from any set of postulates. The question whether an organism can be completely understood in terms of anything simpler than the whole entity can be answered only in terms of an intellectual framework not given in objective scientific data. This framework is implicit in human consciousness and would be available whether or not modern science existed. In fact, long before the sixteenth century men had judged that animals are irreducible wholes on evidence which remains valid for those with eyes to see it. Those "eyes" are given one when one accepts the correct intellectual framework: the "horizon" of being, truth, beauty, and goodness given in consciousness. Harris finds in the data of science evidence for what he already knew anyway. This is an important and worthwhile thing for it makes one's picture of the universe more all-embracing and beautiful, but it is quite a different thing from founding one's philosophy in science.

11. MIND AND COMPUTERS

The ancient philosophical problem about the relationship between the different levels of being in the natural hierarchy assumes different forms in each age. In ours, it often emerges in the question about the relationship between human and "artificial" intelligence. Mortimer Adler discusses this in his excellent work *The Difference of Man and the Difference It Makes*.

According to Adler, there are three possible answers to the question of the way in which two entities differ. They may differ in degree; they may differ in kind, but only superficially; or they may differ radically in kind.

Two entities, say X and Y, differ in degree "(1) where both X and Y have the property *alpha*, and X has less of it, Y more, (2) and where an infinite number of Zs are possible between X and Y, the alpha of each being more than the alpha of X and less than the alpha of Y." [26] An example would be two straight-line segments of unequal length. Sometimes the possible Zs do not actually exist, and then it is easy to think that the difference is more than one of degree. Such, however, is not the case.

A difference in kind exists between two entities when "(1) One of the objects compared possesses a defining characteristic *not possessed by the other*—three-sidedness or four-sidedness, in the case of triangles and quadrangles; divisibility by two or indivisibility by two, in the case of even and odd numbers. (2) There is no intermediate object possible—nothing which is a little more than three-sided or a little less than four-sided, nothing which is somehow in between odd and even numbers." [27]

A difference in kind may be superficial, as in the case of

ice and water, or radical, as in the case of angels and men. Ice
and water are different in kind because one exhibits a stable
crystalline structure, the other does not. However, this discontinuity in a manifest property arises from a difference in
degree at a deeper level. There is a property, the average energy
of the molecules, which falls along a continuum. This continuum has associated with it a "critical point" or "threshold"
such that sets of molecules whose energy is below the critical
point exhibit solid properties and those whose energy is above
it exhibit liquid ones. Because of the presence of the critical
point on the continuous spectrum of the lower level, discontinuity appears on the upper. In general, a difference in kind
is superficial if the discontinuity in the defining property results from a difference in degree in a deeper property, with one
degree falling below, the other above, a critical point.

In the case of angels and men, the situation is otherwise.
Men are dependent on material substructures in a way in
which angels (if they exist) are not. This difference arises
from a difference in kind in their ultimate ground of operation,
or "nature." Hence angels and men are radically different in
kind.

Adler argues that the differences between lower levels of the
natural hierarchy are superficial differences in kind, but that
the difference between man and the other levels of the hierarchy may be radical in kind. He makes use of both the
"negative" and the "positive" edges of Ockham's razor. Using
the negative edge, he shows that the performances of animals
do not require the postulation of any theoretical entities beyond
memory-images and perceptual abstractions, which are totally
dependent on the nervous system and its data-processing
activities. On the other hand, the human performance of propositional speech requires, in conformity with the positive edge
of Ockham's razor, the postulation of "concepts," or "meanings,"
entities which may not be totally dependent on the nervous
system. In order to understand them, it may be necessary to
postulate an immaterial factor in man, something which is in
no way possessed by other animals. If so, man would differ in
kind from other animals at the deepest level, that is to say, he
would differ from them radically in kind.

Persuaded by the traditional Aristotelian–Thomistic arguments for the immateriality of man's power of conceptual thought, Adler leans toward the position that man differs radically in kind from all other animals. He finds this argument suasive rather than apodictic, however, and looks to the future for a final resolution of the question. He believes that if there is nothing in man which is independent of matter, then it should be possible to construct machines having the power of conceptual thought. He discusses this in terms of the "game" proposed by the English computer expert, A. M. Turing. Turing's game involves three players, two of them human, one a machine. One of the human players is the interrogator who proposes questions to the other two via teletype. They reply by the same means so that none of the players ever sees another. The object of the human players is to determine which of the respondents is the computer; the object of the computer, of course, is to conceal this. The game is simply an imaginative way of focusing on the question, can a machine be constructed which possesses the power of conceptual thought and is consequently able to use propositional language? If this could be done, it would not be possible, under the limited conditions of the game, to distinguish such a machine from a human being.

Adler argues, rightly I think, that if man is not radically different in kind from other natural beings, then it should be possible to construct such a machine. It is clear that the attempt will be made in the future, indeed, that computer experts are already laying the groundwork for such an attempt. Consequently, Adler expects that at some future date either the success of a robot in Turing's game or a series of repeated failures will decide the issue about the nature of man. Construction of a thinking robot will show that there is nothing immaterial in man; repeated failure to do so will make it highly probable that there is an immaterial factor in his makeup.

Adler's interesting work raises a number of issues which are worth discussing. First of all, there are certain difficulties about the concept of a superficial difference in kind. This type of difference is said to exist between two entities which do indeed differ in kind, but in such a way that the difference is the result of an underlying difference of degree along a continuum which has a critical point. One says that two entities differ in kind when one possesses a *defining* characteristic which the other does not have. A definition can be essential or merely descriptive. If the definition is essential, the defining properties which comprise it are fundamental constitutive notes of the essence of the being in question. It is difficult to see how such a property can be due to a deeper property when it itself is already on the deepest level. If, on the other hand, the definition is merely descriptive, the defining characteristics which comprise it are not descriptive of the inmost reality of the entity in question. It would not be very significant that two entities should differ in kind if all that a difference in kind means is a difference with respect to a superficial characteristic. Consequently, the utility of the concept of a superficial difference in kind seems open to question.

Applying the concept to man is especially awkward. In this case, the defining characteristic which man has and other entities do not is the power of conceptual thought, which is equivalent in Adler's view to the ability to do science, create art, formulate and follow moral ideals, love in a disinterested way. The underlying structure from which these abilities supposedly arise totally is a complex nervous system. This property does not exist on the personal level; it is a matter of complexity of physical structure rather than of insight and free decision. Accepting the idea that man might differ only superficially in kind from the higher animals means accepting the possibility that the personal activities of understanding and love may not be particularly noble or significant. This acceptance is much against Adler's inclinations, but he is forced into it by his emphasis on the objective and the measurable and his neglect of human subjectivity and implicit knowledge. As a result he does not see the full force of the fact that knowledge and love are defining properties of man in a very strong way: they define him essentially and are what is deepest in him. There simply is nothing deeper in the material universe; thus, there is no possibility of an underlying continuum with an associated critical point such as is necessary to ground a superficial difference in kind.

The same neglect of the subjective is responsible for the fact

that Adler thinks that animals differ only superficially in kind
from plants, and plants only superficially in kind from inorganic nature. Animal knowledge and desire, if seen in the
light of human experience, appear as the integral act of a
sentient being with its own kind of subjectivity. The ability
to perform such acts is an essential defining property which
cannot depend on anything deeper within the animal. This
ability is the deepest thing about the animal, and the acts of
the animal's nervous system depend on it, insofar as the animal
is different from an inorganic system.

This must not be interpreted as a denial of the very evident
fact that higher activities depend on lower ones as conditions
of possibility. But what is higher and distinctive about a
system cannot be reduced to the lower level.

In the end, a superficial distinction in kind reduces to being
an apparent distinction in kind which is really a distinction in
degree. As soon as one recognizes that a defining property is
dependent on a deeper property with respect to which there is
no discontinuity between the two entities being considered, one
also realizes that one has chosen one's defining property badly
and that it should be eliminated from the definition if one
wants to have anything more than a mere descriptive definition
which does not exhibit very perspicuously the nature of the
entity under consideration.

The notion of a superficial difference in kind is really not
very useful, except for reductionists who want to admit the
evident fact that the ability to know and love is man's most
important characteristic, and yet at the same time reduce this
ability to less important properties. Reductionists thus find
themselves in a very awkward position: they deny that they
are saying that man is "nothing but" organized inorganic matter,
and yet they give no hint, not even to the extent of a negative
definition such as "immateriality," of what the "something
more" that he has could be. This would logically lead to a
denial of the principle of causality in the case of this "something more" or to a denial that what is intelligible must be
capable of being understood, at least in principle.

There is no problem about saying that man is "something
more" but that we do not understand exactly what this means.

Indeed, those who assert the existence of an immaterial principle in man are doing little more than this. They are saying that there is something more in man's behavior than material structure can account for and that this is the result of something more in his being. The truth of this very minimal claim would be obvious were it not for the ingrained presupposition of the modern mind that everything must fit into the Procrustean Bed of scientific explanation.

What about the truth of Adler's claim that the construction of a "machine" capable of conceptual thought would prove that man's distinction from the animals is a superficial distinction in kind? Adler himself notes that this event would not prove the point definitively since one might claim that the new entity itself contains an immaterial principle. However, he dismisses this argument as implausible—overhastily, in my judgment.

One must distinguish between the manner of construction of a system and its ontological rank. If one examines the structure of the arguments which conclude that there is an immaterial principle in man, one finds that they are not based on the mode of his origin but on his behavior as an existing being. Hence, they apply unchanged to a being which exhibits the same behavior, even if its mode of origin may be different.

Neither is there anything self-contradictory in the notion of synthesizing a living system, nor in the notion of synthesizing a thinking system. After all, the ecosystem of the earth, at a time when it contained no intelligent life, produced man. There is no obvious reason why that same system, now that it has become more highly developed through the emergence of man, cannot through his agency produce still other intelligent beings. Indeed it does so in the conception (or the hominization, if that is a distinct event) of each child. What would be self-contradictory would be the notion that a material system, even one containing partly spiritual beings such as man, could be the total cause of intelligent life. As we shall see later, nothing less than the cooperation of God is necessary for that.

However, I am quite skeptical about the possibility of a rational entity which is not at the same time sentient and vegetative or a sentient one which is not also vegetative. The patterns which matter must assume in order to be intelligent or sentient are determined by the overall structure of the cosmos. We do not know even our own pattern in complete detail, and we cannot rule out the possibility that some configuration of semiconductors or thin films might do the trick, but it seems unlikely to me. I would expect that an intelligent being must occupy a niche in the hierarchy of natural systems and be produced by evolution. Life is a dynamic state in which the interaction of the parts transforms them inwardly and raises them to a new level. Consequently, a "part" does not have the right properties until it is actually in the living system. I suspect that the only way to synthesize sentient or rational life is to follow in nature's footsteps by beginning with a very simple organic system and gradually complexifying it. Perhaps the science of the future will be able to reduce considerably the three or four billion years it took nature to produce the higher animals, or possibly to improve on the materials she used to construct some of their parts, but I doubt if the time can be reduced enough, or sufficient improvement achieved, to make the job worthwhile. This, of course, says nothing against the possibility of making comparatively minor alterations in existing forms of life or of synthesizing simple forms *de novo*.

Although Adler dismisses too quickly the possibility that man (as a partial cause) might synthesize an entity which would be partly constituted by an immaterial principle, his reaction is probably inspired by a valid insight. If a machine were constructed of macroscopic parts so that the action of each individual part could be observed, and if it were found that the activity going on was exactly what would be predicted by the laws of inanimate matter, then it would indeed be clear that the total machine did not contain an immaterial principle. Even though the total action of the machine might not be actually predictable, because of its complexity, it would be predictable in principle, just as the action of a complex power grid is predictable in principle even though it is not predictable in practice, as "blackouts" sometimes testify. If such a machine were to possess the power of conceptual thought, this would indeed falsify the proposition that an entity must possess an immaterial principle in order to be capable of conceptual thought.

However, a machine of this type is an unlikely candidate for intelligence, even from the viewpoint of a materialist. In order to match the complexity of the human nervous system the robot makers will have to make use of miniaturization, very likely in a much higher degree than we have yet seen. As the parts and their actions diminish in magnitude, quantum phenomena and the uncertainty principle will play an increasing role. In all probability, if an intelligent entity is ever synthesized, we shall find it just as hard to measure and predict what goes on inside it as we do to measure and predict what goes on inside a man. It is unlikely that science or technology will ever provide clear-cut decisions to fundamental philosophical or theological options. In fact, I do not doubt that if intelligent "machines" are constructed someday, people will ask them whether they believe in God and whether they think they have immortal souls, and that the replies given will cover the same range as ours. If they are more intelligent, they will support the same basic positions with more subtlety, but the same arguments between atheists and theists will go on among the "machines."

There seems to be a "theological uncertainty principle" operating here. Heisenberg once noticed that nature was apparently involved in a conspiracy to prevent us from ever getting simultaneous, accurate measurements of the position and momentum of elementary particles. From this observation he drew certain conclusions about the nature of physical reality. Similarly, it seems that reality is engaged in a conspiracy to prevent us from deciding fundamental questions about the nature of man, the existence of God, etc., on the basis of objective evidence alone. Subjective attitudes and implicit knowledge always play a paramount role, and it seems that this is designedly and necessarily so.

12. THE HOLISM OF NONLIVING SYSTEMS

We turn now to the consideration of systems below the level of life. Does the same holistic viewpoint needed to understand life apply to these systems as well? The desire for theoretical coherence suggests that it does. The hierarchy of natural systems is an expression of something deep in the nature of the cosmos, and presumably this same factor operated in the formation of the simpler as well as of the more complex systems.

Consider the importance of interactions in present-day physics. It is now known that the so-called elementary particles transform into one another in complicated ways, and it is becoming increasingly difficult to use them as stable conceptual pegs on which to hang one's ideas. The most stable and invariant features of matter seem to be the various interactions. But interactions go on within a system. No one can deduce how particles will interact by considering their behavior as free particles. The physicist has to let them form a system by interacting and then abstract the nature of the interaction and the physical symmetries associated with it from the system. In other words, interactions are emergent properties which appear only when a new system is formed. The properties known from studying the parts as individual entities do not predict the interactions.

Quantum field theory is still the most fundamental theory which physics knows, in spite of its inconsistencies. It is far from clear that this theory accords with reductionism. A field is conceived of as a single entity which manifests itself in various regions of space–time by means of its characteristic excitations or particles. The whole set of identical particles associated with a single field shows correlated behavior. Thus, in the case of electrons and other particles of spin ½, no two particles may be in exactly the same state. This principle, the Pauli exclusion principle, is responsible for the architecture of the atoms, which builds up in a systematic way which would not be possible unless all electrons were part of a single system which imposes constraints on its parts.

As a matter of fact, even the foundations of classical physics demonstrate that reductionism is not an ultimate viewpoint. Matter in all parts of the universe obeys Newton's gravitational formula to a very good approximation. Why should particles separated by enormous gulfs of space and time have similar behavior? Clearly, something outside the individual entity is needed to explain this. The various individuals must be part of a larger system which determines the nature of the parts. The differential equations of Newtonian physics are not disembodied entities hanging in some Platonic realm of ideas. Neither can they be determined entirely by the nature of individual particles, for this would not explain their universal applicability. They are abstracted from the concrete behavior of the coherent system we call the universe.

It is highly doubtful therefore that reductionism applies even to inanimate matter. Even the simplest natural physical systems are too subtle to be understood completely in a reductionistic way. However, a reductionistic approach yields a more complete explanation about an atom than it does about a man. The individuality of simpler systems is not as marked as that of the more complex ones. To de-atomize an atom does not do it much of an injustice, but to dehumanize a man is an atrocity.

NOTES

1. See BERTALANFFY and BUCKLEY for introductions to this field.
2. POLANYI (I), pp. 176–177, 328–332.
3. Polanyi seems to overlook this distinction in his discussion of machines.
4. Von Bertalanffy, L., "General System Theory—A Critical Review," in BUCKLEY, pp. 17, 21.
5. Strangely (or, perhaps appropriately) enough, it is often physicists who show the greatest awareness of this. Thus Brillouin has remarked that it is quite possible that, far from being "reduced" to chemistry and physics, biological phenomena may demand a revolution in physics. See JAKI (I), p. 327 and BUCKLEY, pp. 147, 148.
6. See POLANYI (III), p. 40.
7. Clearly, we are defining "subjective" and "objective" primarily in terms of the relational reality of knowing and not in terms of verifiability or of any other criterion.
8. WHITEHEAD (II), pp. 72ff.
9. See the works of Polanyi listed in the bibliography. GRENE provides a good introduction to Polanyi's thought.
10. KUHN, especially Chapter 10.
11. See BARBOUR, p. 308.
12. LONERGAN, especially pp. 205, 206.
13. DeWITT (I), p. 30. See also DeWITT (II). For an explanation of Heisenberg's position, see HEISENBERG (I) and (II) and HEELAN.
14. POLANYI (I), Chapter 12, and part 4.
15. HARRIS, especially part II; also POLANYI (II).
16. PERLS contains many good insights about this.
17. Besides the work of Piaget, see HEBB, TINBERGEN, and HARRIS who has an extensive bibliography.
18. KELLOGG, cited by POLANYI (I), p. 69.

19. The analogy comes from modern communication theory. See SHANNON.

20. These ideas are developed in greater detail later.

21. LORENZ (I), p. 235; cited in POLANYI (I), pp. 243–244.

22. TINBERGEN, pp. 2–5.

23. TINBERGEN, p. 4.

24. HARRIS is a good starting-point and contains extensive bibliographical references.

25. See CORETH for an exposition of this position.

26. ADLER, p. 30.

27. ADLER, p. 19.

2

Evolution

1. HOLISM AND EVOLUTION

THE ENTIRE NATURAL HIERARCHY, then, is characterized by holism. This is certain in the case of man, probable in the case of nonliving matter. This remarkable feature of the present-day universe must be related to the nature of the evolutionary process which has built it up. We know that the present-day structures of life have evolved from non-living matter through a sequence of forms of gradually increasing complexity. It is not, of course, individual living entities which have changed their form, but rather populations of living entities and, ultimately, the entire planet-wide ecosystem. In fact, it is clear that the entire universe is involved in the process of evolution. Biological evolution here on earth is only the latest chapter in a history which probably involved the expansion of the universe from an enormously dense initial state, the formation of stable elementary particles, the condensation of stars and galaxies, the synthesis of the heavier elements, and finally, the formation of planets, like the earth,[1] which are suitable for life.

At this point we should distinguish between two kinds of evolution. The first results in change, but the final product exists at the same level of organization as the original entities

41

with which the process began. The total evolving system has not advanced to a higher level. Such, for example, would be the motion of an open chemical system toward equilibrium, the muting of the colors of butterflies in regions of industrial air pollution, the adaptation of a species of insects to the presence of DDT. The second kind of evolution we can call "emergent." Here the process results in the ascent of the total evolving system to a higher level. At the end it contains entities of greater value than it had at the beginning. This increased value implies that some new and higher kind of organismic regulation has also emerged to control the coordination of the lower-level entities into the new superordinate ones.

Our holistic viewpoint which requires us to assert that the universe has indeed ascended to higher and higher levels of being and value confronts us, therefore, with a problem of "emergence." How can we account for the fact that a world of complex molecules transformed itself into a world of life, then of sentient life, and finally of intelligent life? The greater cannot emerge from the lesser, at least not by the sole power of the lesser. To assert this is to deny the principle of causality. The excess in being and value of the later stage over the earlier would emerge from nowhere and be attributable to no cause. But if being and value can come into existence without adequate cause, then anything at all could come from anything else, or from sheer nothingness, at any time.

The reductionist does not have to deal with this problem. If higher forms are completely explicable in terms of lower ones, then there is no excess of value or being at a later stage and nothing which requires explanation. Neither is there any real progress or variety, just a factually different state of affairs which is "better" from our human point of view, but which in reality has no more absolute value than the primitive state of complete molecular chaos. Once we reject reductionism, however, this easy and dehumanizing escape is closed to us. We must conclude that the universe as a whole, which is both subject and agent of evolution, already precontains in some way the value and being which become manifest in a different form with the evolution of sentient life.

The dynamism which powers the development of the universe has a formally intelligible pattern. Man has already glimpsed parts and aspects of it in the "laws" of physical and biological science. It is conceivable that we may someday understand it much more fully and give partial expression to our understanding in a unified mathematical way. Nevertheless this dynamism is ultimately more like a desire than a mathematical equation. It has a formal aspect, but above all it is a power which aims at being and value. We ourselves, inasmuch as we are a part of the universe, feel this power, this desire, within us. Since all activity and all development spring ultimately from desire, every agent, insofar as it is an agent, acts for an end. This applies to the cosmos as a whole. Evolution is a teleological process which moves toward a goal.

2. NEO-DARWINISM

In discussing evolution, one has to distinguish between philosophical and scientific theories. To some extent this is a distinction between ideal, rather than real, entities. It is difficult to give an account of evolution without taking a stand, either deliberately or inadvertently, on philosophical issues. Unfortunately many modern accounts of evolution which set out to give a purely scientific, and therefore partial, account of the phenomenon are tainted with reductionism and anti-teleological bias. I shall call this kind of modern reductionistic evolutionary theory "Neo-Darwinism." I shall argue that it is manifestly incorrect and that an adequate explanation of evolution, which takes into account evidence about the nature of man readily available to everyone through personal experience, must include teleological notions. This does not, of course, preclude a partial explanation within the realm of biological science which leaves unsolved questions about finality and about the nature of man which biology does not aim to settle.

Contemporary theory holds that two fundamental processes advance evolution: the production of variation among the members of a population of organisms and the selection of certain variants by the environment. Natural selection operates on the total organism in terms of its "phenotype," the totality of its characteristics, including behavior as well as physical structure. Variation of phenotype within a population has many causes, but the variation having the most evolutionary significance occurs in the genetic endowment, the "genotype." This is the result of both recombination and mutation, but mutation is the root cause of genetic, and, therefore, of phenotypic, variation. Hence we shall focus our discussion on the two basic concepts of mutation and natural selection.

Neo-Darwinism regards mutations as strictly chance events. They can result from a number of different causes, but in no case do the causes exhibit any systematic pattern beyond what would be expected from the physical and chemical nature of the environment. Naturally there will be more mutations if there is an increase in radioactivity, and if this radioactivity is of a particular type, then certain chemical bonds may be affected more than others. But within the limits set by physical and chemical constraints, the mutations which occur are strictly random.

Natural selection is not a random process, but it too is considered nonteleological. Those organisms which survive and reproduce do so not because they are better in any absolute sense, but simply because they fit in better with the environment and happen to have a physical form and behavioral tendencies which lead to survival.

If a particular mutated gene gives rise to a bodily form and to a type of behavior such that the animal possessing it has a better-than-average chance to survive and reproduce, then the laws of probability show that this gene will tend to spread throughout the population.[2] If the gene has effects unfavorable to survival, then it will tend to die out. The rate of mutation is such that in the course of time the population will tend to diversify into all possible forms which have an ecological niche available to them. In other words, any way of life which can fit in with the ecosystem will eventually be assumed by some species of animal. The process is somewhat akin to the diversification of business activities within a free enterprise system. Some company will eventually take up the manufacture of *any* product for which there is a demand including even dope, prostitution, etc.

According to this view, success in surviving and reproducing is the decisive factor in determining what sorts of animals will possess the earth. This success is just a factual thing, and nature knows no "oughts." If men exterminate themselves and leave the earth to the insects, this will be an indication that intelligence was an unsuccessful experiment, and there will be no real loss to the universe, no matter how much we might be tempted by emotional factors to think so.

In this type of evolutionary theory, we see reductionism at its best and at its worst. On the one hand a few basic and simple ideas lead to a unified explanation of a vast range of important phenomena. Although this explanation involves a number of unproven assumptions, it cannot be denied that the theory has a certain elegance and grandeur. But on the other hand, the cogency and beauty of the theory depend on the suppression of human values and of certain knowledge about the nature of reality which can be obtained by reflection on human experience. Every man knows implicitly in the very performance of the actions that knowledge and love are "higher," "more valuable," "more real" than any activity of brute matter. A man is absolutely more valuable and of higher ontological rank than a chimpanzee or a roach, and any theory which contradicts or minimizes the importance of this fact cannot be correct.

It is clear that no one can prove a reductionistic theory of evolution in any rigorous sense. Such a proof would require one to know the precise structure of the genetic material existing at each stage of evolution, the probabilities of the various possible mutations, the effects of these mutations (singly and in combination) on phenotype and behavior, and finally the way in which changes in phenotype and behavior interact in the ecosystem to produce natural selection. Clearly enough, a precise understanding of these matters is out of reach for the foreseeable future.

Furthermore, even if one is willing to accept a considerably less than rigorous proof, there is reason to be dissatisfied with present Neo-Darwinian theory. It is clear that it is very incomplete.[3] The building blocks of living tissue are the long chainlike molecules called proteins. The links in the protein chains are the amino acids, of which twenty different kinds are used. If we limit our consideration to chains of length 250 or less, this gives about 10^{325} abstract possibilities for making proteins. The number of protein molecules which have ever existed on earth is probably considerably less than 10^{52}, a number which is almost infinitesimal when compared with the first one.[4] Yet among this relatively tiny number of possibilities are all the proteins which nature has found useful in constructing living organisms. One is very reluctant to say simply that nature hit the jackpot here on earth in the face of tremendous odds and that that is all there is to it. Several possible explanations suggest themselves: (1) Useful proteins are common among the possible ones. This possibility cannot be ruled out, but there seems to be some evidence against it.[5] (2) Nature's search for useful proteins has been guided by some very strong constraints. (3) Nature discovered a useful protein very early in the history of planet Earth, and it happens that among those located close to it in the space of protein possibilities (i.e., among those which can be obtained from the first by a relatively small number of changes), there are relatively many useful kinds. This third explanation is related to the second. The strong constraints of (2) would be the starting-point of nature's search and the topology of the protein space, that is, the way in which useful proteins are related to one another. The probability of this third explanation is enhanced somewhat by recent evidence that complex organic molecules may be formed in the primeval dust clouds from which solar systems condense.[6] Thus the search for a good starting-point would be carried on by nature over the vast extent of space before the planets are formed. The universe may be only three times as old as life on earth, so the time factor is not very important, but space is vast and in such a large laboratory nature might have a good chance of concocting almost anything.

Any of the foregoing explanations might be correct—but again they might not. In any event, to have a complete theory, it will be necessary to be far more specific and to produce evidence for what is specified.

There are other complications.[7] According to current theory, cells construct protein under the control of information stored in the structure of the genes in the nuclei of the cells. Mutations of the genes result in changes in the proteins, which are then selected by the environment. Genes, or, more properly, cistrons, and proteins correspond in a one-to-one way, but the topology of the gene and protein spaces does not necessarily correspond. What that means is this: As far as the set of genes is concerned, the relevant meaning of "closeness" refers to the ease with which one can be derived from another by random changes. As far as the space of proteins is concerned, the relevant meaning of "closeness" refers to the changes in the "fitness" of the organism constructed from the proteins to deal with the environment. There is no *a priori* reason why the two topologies (or networks of closeness relations) should not be wildly different; if they are, evolution would be blocked. This occurs in computer experiments in which the instructions stored in the machine are the genetic information and the computations performed are the behavior of the system. A slight change in the instructions usually produces nonsense in the computations, and a slight change in the computations requires extensive changes in the instructions. So far, no detailed explanation has been given for what seems the evident fact that the two topologies do as a matter of fact correspond rather closely (according to some relevant definition of "closely").

It is clear then that current Neo-Darwinian theory is very incomplete. What can we reasonably anticipate about the nature of the factors which will have to be added to it in order to complete it? Will they be essentially subordinate to the principal concepts of random mutation and natural selection, or will they be radically new concepts which are just as important as those already included in the theory? At the 1966 Wistar Institute symposium on "Mathematical Challenges to the Neo-Darwinian Interpretation of Evolution," the theoretical physicist Victor Weisskopf explained his feeling about Neo-Darwinism in terms of a comparison between the quantum mechanics of atoms and the quantum mechanics of nuclei.[8] Physicists feel that they already possess all the basic concepts needed to explain atomic phenomena in spite of the fact that they are unable to deduce the behavior of any atom except hydrogen from first principles. Their success with hydrogen gives them confidence in their theories, and as far as can be ascertained, only mathematical complications prevent the same complete success with other atoms. Nuclear physicists, in spite of their considerable success in explaining some phenomena, feel that a new fundamental principle is probably needed. The nature of the unexplained phenomena is not such as to engender confidence that more detailed, accurate, and lengthy computations based on current theory will yield success. Weisskopf's feeling about Neo-Darwinism is that it is similar to current nuclear theory. The missing postulates are not likely to be essentially subordinate and minor ones. Other leading physicists have expressed similar views.

The mathematicians and physicists at the above-cited Wistar symposium appear to have been notably more negative toward Neo-Darwinism than the biologists. Why? As Polanyi has pointed out, the role of implicit knowledge is greater in biology than in physics and, *a fortiori*, mathematics. Physicists are less rigorous than mathematicians, but even they are used to making more rigorous and explicit deductions from their postulates than are biologists. If the results agree with the prediction, there is reasonable assurance that all the relevant factors have been caught by the web of definitions and postulates. If they do not agree, or if the deduction cannot be made even for the simplest case in a class of problems, then one tends to think that the set of postulates is incomplete or wrong.

Biologists, on the other hand, because of the extreme complexity of their subject, are seldom able to make detailed predictions. More often than physicists, they must be content with making general predictions subject to many exceptions or with explaining things after the event. In doing this they quite rightly rely on implicit knowledge and are influenced in their tracing of causes by the already-known outcome. But there is danger in this. It has been pointed out that current evolutionary theory is not easily falsifiable.[9] The way in which the postulates are used is such that an explanation will be forthcoming no matter what happens, and directly opposite conclusions are sometimes explained with equal ease.[10] This is not surprising in a teleological theory in which events are ordered to an end which can only be fully known when the final higher-order pattern has come into existence, and in which the laws obtaining on the lower levels are only partial determinants of the results. As far as the action of lower-level entities is concerned, there is a certain amount of indeterminacy, and the final result can go either way depending on the nature of the higher-order pattern which is coming into existence. But in a professedly nonteleological theory in which already-existing lower-level structures determine the whole course of events, it seems strange not to be able to distinguish more sharply between outcomes which would support the theory and those which would falsify it. It seems possible—in fact, from within my teleological framework it is evident—that biologists, trained as they are to grasp the total pattern of biological functioning and to see the workings of the parts in terms of this pattern, are unconsciously supplementing the reductionistic postulates of Neo-Darwinism with their implicit understanding of the holistic character of life and its operations.

Certainly I have no objection to implicit knowledge or to unfalsifiable theories. One can arrive at the most important truths, such as the unique dignity of man or the existence of God, only by making use of implicits. Again, the theory that God has created and now rules the world is unfalsifiable. But the unfalsifiability of these notions is the result of the transcendent, infinitely rich character of their content. Neo-Darwinism, on the other hand, is unfalsifiable by reason of its poverty and the vagueness with which it is used. The factors of mutation and selection almost certainly have something to do with what happens, but because no one can specify exactly what that something is, one is free to suppose that they do just what is required. But one is also free to suppose that they do not do just what is required. There is no justification for claiming that one has an even relatively complete nonteleological theory when there is no possibility of deducing the consequences of the postulates even in an approximate way. At the Wistar symposium Weisskopf remarked: "If I wanted to be nasty toward the evolutionists, I would say that they are surer of themselves than we nuclear physicists are—and that's quite a lot." [11]

It must be noted that it is also impossible to give a scientific proof of a holistic theory of evolution. One can argue that if there is no organismic self-regulation of the details of life-processes by the whole organism, then the number of mutations producing bodily configurations and behavior which will work successfully must be very small in comparison with the number which are possible on the basis of physical and chemical laws. In this case random mutations followed by natural selection would be very unlikely to produce an organism as complicated and well adapted as man is in the three or four billion years in which life has existed on earth. Arguments of this kind suffer, however, from the same problems as the reductionistic ones. No one knows enough about the probability of the various biochemical states and transitions involved in life to assign probabilities to them with any degree of certainty. The same can be said in general about all claims that a given vital process could not possibly be explained on the basis of mechanism alone. These arguments are persuasive only to those who, like me, are already convinced. One can never be sure that some new fact or theory may not be forthcoming which will give those so inclined a more comfortable way of retaining their reductionism. At the present time, and most probably at all times, the question of reductionism versus holism cannot be decided on strictly scientific grounds. There are simply two schools of thought, both equally "scientific." One can decide between them only on the basis of considerations which carry one beyond the bounds of science.

3. TELEOLOGY AND EVOLUTION

In what specific ways should the ordinary account of evolution be altered in order to take account of teleology? In the holistic perspective which we have already adopted, one way at least is clear. An organism lives by constructing and utilizing living machinery through a process of organismic self-regulation which itself is not mechanistic. A mutation in the germ plasm will certainly have effects on the organism which inherits it, but what those effects will be is not totally determined by mechanistic principles. The organismic self-regulation of the total organism must use and develop the substructures given in the genes. Consequently, enhanced performance after a mutation is the result not only of the change in the structure of cell nuclei because of the mutation but also of the manner in which this change is utilized. Since the relationship of various substructures and processes may be reshuffled in various ways by organismic regulation, the chances of getting favorable results from a mutation increase considerably. Thus in a poker game a "wild card" is of considerably more value than an ordinary one because it can be fitted into a number of different patterns.[12]

Another element of the process of evolution which may be influenced by teleology is the occurrence of mutations. From a holistic view of the structure of matter, it is clear that the entire universe is a coherent system which develops in accord with an intelligible pattern immanent in its evolutionary drive. At this level of explanation, events which are random with respect to the laws operative at lower levels can be seen to be intelligible. Hence it is plausible to suppose that nature has managed to find comparatively direct and improbable paths through the space of possible nucleotides to arrive at genetic configurations suitable for higher forms of life.

However, I do not care to assert that this is certainly the case. It may happen that the physico-chemical constraints on motion through the nucleotide-space are so strong that the paths by which nature has proceeded are quite probable ones (probable, that is, even with respect to the physico-chemical level of explanation). In this case there would be no need to suppose the existence of any higher-order constraint intelligible only on the level of the evolutionary drive of the universe.

Here we see displayed the ineffectiveness of holism as a scientific theory. As I have remarked several times before, it is by no means bad science to assume reductionism as a working hypothesis, or rather to assume that the particular phenomenon under investigation can be understood in terms of mechanisms. Perhaps some day scientists will find a way to treat the concept of organismic regulation in a scientific way but as far as I know no one has as yet. At present the only way to proceed seems to be to explain everything which can be explained in terms of mechanisms, and then attribute what is left to organismic regulation. Perhaps it is just as well that there are philosophical reductionists. Their philosophical error may give them the psychological stamina to continue looking for mechanisms when more philosophically enlightened persons withdraw. Columbus' underestimation of the size of the earth (a rather gross blunder, one must admit) turned out to be quite fortunate.

However, human society can accommodate only a certain number of reductionists, for their intellectual productions erode consensus about central human values without which no society can remain sane. One can draw an analogy with the sickle-cell-anemia gene. A certain number of these genes is useful to a population which must live in a malaria-infested area without the benefits of modern medicine. However, too many of them is worse than the malaria.

4. TIME

The process of evolution, which manifests itself most clearly in biology, geology, and astronomy, reveals something about the fundamental physical concept of time which could not easily be learned in physics. The nature of time has puzzled many philosophers. As St. Augustine said, we understand it implicitly well enough to live our lives in time without difficulty but when we try to explicitate what we understand we are at a loss.[13] It is easy to fall into all sorts of paradoxes which are not easy to resolve. But the notion of development, acquired from consideration of evolution, and also from our own personal experience, provides a key to understanding.

We are aware of ourselves as substantial beings whose basic performance of being perdures and develops through a sequence of states. I am now sitting here typing, aware that I am possessed of a certain amount of knowledge and experience and that I have lived through a history of a certain length. Yesterday my history was a day shorter, and the knowledge and experience I have acquired during the past twenty-four hours did not exist. I have developed during this time and am now more than I was then—in terms of experience at least. I have good reason to hope that tomorrow I shall still be alive and that I shall be able to look back at today in a similar way, and I am straining forward toward the realization of my various goals tomorrow and in the days to come. But right now tomorrow does not exist; it is a mere anticipation contained in my thoughts and desires. Thus there is an obvious asymmetry to our experience of time. The future is a mere expectation, but the past has a certain consistency. What has happened can never be changed, and in its refractoriness to our desires it has a certain objectivity. And in fact our personal acts of the past endure within us. Much of what we thought is accessible to us as memory, and much of what we desired and decided continues to influence us right now.

But just how does the past endure within us? This is equivalent to asking what is the truth of history. What grounds the objective truth of the statement that a man named Kepler discovered the laws of planetary motion near the beginning of the seventeenth century? St. Augustine believed that there are three times, the past, the present, and the future. For him only the present exists in the fullest sense. The future is a mere anticipation, and the past is contained only in memory, a present reality.[14] This was also the theory of "Big Brother" in Orwell's novel *Nineteen Eighty-Four*. Big Brother felt that if the past continues to exist only in the structures it has produced in the present, then it should be possible to change the truth of history by simply revising those structures. Of course it must be admitted that it is beyond human power to do a complete job of changing the traces of the past, but if enough were done along those lines to make it impossible to reconstruct the past with certainty, then it would seem that the past would no longer have any determinate nature at all and that one opinion about it would really be as good as another. And, indeed, if God wanted to play this game, He could succeed perfectly in changing history by simply revising the present.

There seems to be something wrong with this point of view. We judge intuitively that history has an objective truth which cannot be tampered with. What has happened, has happened, no matter how many records of the past are changed, and no matter how much God or man may wish otherwise. What are the implicits which are latent in this intuition? At the root of our understanding of historical truth is our understanding of being and change. The reality of the present does not simply cease to exist when it becomes part of the past. It is integrated into the subjective reality of the agents which produced it. A senile old person has lost the power to act and to manifest his inner self as he did in the days of his youth, but somehow he is more even in his ruin than he was as a youth. He has lived, experienced, suffered, and in some mysterious way there is more to him now than there was in the days when he could perceive and think so much more intensely. That which once comes into being is somehow everlasting and beyond the succession of time insofar as it exists. Change is not, as Aristotle thought, the replacement of one principle of being (accidental or substantial form) by another, but rather the development of an enduring reality into something greater which still includes what was there to start with. If one refuses this notion of change as development, there is no way of avoiding Aristotle's logic and accepting his ultimately unsatisfactory notions of accidental and substantial change. One must then explain real being and real change in terms of "principles of being" (substance and accident, matter and form) which themselves are not fully real but which nevertheless are prior to real things in the order of explanation. This means that there is a lack of balance between the orders of being and of explanation. In the order of being, real things are primary, but in the order of explanation they are not. It cannot be demonstrated that the Aristotelian view is logically contradictory, but I have concluded that it is implausible and is not an adequate expression of my subjective and implicit understanding of change.

Rather, change is development. The past exists within the present and is in fact the cause of the new reality which is now existing for the first time. The Augustinian view of past times requires one to suppose that the cause of the present (i.e., the past) ceases to exist before what it causes can come to be, or else to suppose that God and not the past is the cause of the present. The first supposition is unsatisfactory because it denies our basic intuition that cause and effect must exist together in the unity of a real relationship, which is impossible if one term does not exist. The second is a kind of occasionalism which has God causing everything on the basis of His knowledge of the no-longer-existing past and denies real causality to finite being.

I suggest that the development of the entire cosmos must be understood by analogy with human experience. At each instant of time, the past brings forth a new increment of reality which we can call the "objective present." The entire existing past is on the originative, subjective side of the action. The new increment of reality is on the terminative, objective side. As each objective present becomes a part of the past, it is integrated into the subjective reality of the agents which brought it forth. On the highest level of the hierarchy, this means that the objective present of the entire cosmos gets integrated into the subjective reality of cosmic history. The act by which the cosmos brings forth a new objective present is analogous to a human decision. The relationship between the status of an event when it is present and when it is part of the past is the same as that between a decision now being made and that same decision when it is past. A quality of objective immediacy is lost, but the decision continues to exist and to produce effects.[15]

Thus the cosmos is not a three-dimensional structure but a four-dimensional one growing in time. The basic image for the Augustinian view of time is that of a spot of light moving along a line. The only point of the line which is "really real" is that on which the light falls now. Once the light has passed, the point it has illuminated falls back into a state of unreality. The image I am suggesting is that of a building which is being erected. Workmen are now busy putting up the ten-billion-and-first story, which thus has a special status. But the first ten billion have not ceased to exist. Indeed it is their continuing reality which undergirds and makes possible the activity which is going on at the top.

This understanding of time casts light on the nature of the evolutionary process. The reality of the universe is continually increasing in richness as time adds successive increments of structure to it. Since the activities which take place within it spring from an ever richer and more complex ground, it is not surprising that these activities are ever richer and more complex. The "laws" which govern elementary particles may be the same now as they were ten billion years ago (of course they may also not be the same), but they are now supplemented by additional laws which govern the activity of living systems. Intelligent life could appear only when the universe had reached a certain maturity. The universe bears its history within itself as the source from which its present state emanates. What has been achieved can never be lost or fail to find some form of expression in the restricted present, except, perhaps, because of the deliberately perverse decisions of rational beings.

In the case of man his entire being, which includes all his past, determines his present situation. Noble deeds and ignoble ones alike are a part of our personality forever and exert an influence on our present behavior; these deeds may be counteracted or even used in a way contrary to their original intention, but they cannot be annihilated.

This view of time is not as common as the Augustinian one because our past does not have the same immediacy as our present. We are aware of it only subjectively. Our objective awareness includes only the new structures now being born. Even when we remember the past our act of awareness does not attain it in itself. Rather, by the subjective power of our already constituted structures we bring forth a symbol, the present memory-image, which is expressive of what we already are, and thus we bring ourselves face to face with our own past. Thus even though the past still exists and can properly be called the "subjective present," it lacks the immediacy of the objective present because it is known subjectively and implicitly rather than objectively.

The subjective richness of our conscious experience is a function of our whole past; its intensity is a function of what we are trying to express in the objective present. As a result, the intensity of consciousness varies from moment to moment and can reach a very low level at times when our self-expression is mainly confined to the vegetative plane of existence.

It might be objected that the view I have developed threatens the traditional view that God conserves creatures in being. If the past continues to exist and if it cannot be changed even by God, then it would seem to be independent of God's

creative action. But I am not asserting this. Rather I say that
the past is just as permanent as God's creative act. Once the
divine decision has been made it simply is, and there is no
possibility of its not being. As a result its term, the finite reality
in question, must endure also.

5. THE BEGINNING OF TIME

The principles we have discussed have a bearing on the question of whether the universe had a beginning in time. If the
passage of time is really the development of the cosmic structure, then either the past is finite in extent or else the qualitative richness of the objective present oscillates up and down
rather than increasing monotonically. For, if it does increase
monotonically and has had infinite time to do so, then the objective present and the activity which produces it would have to
be unlimited in richness and intensity, which is not the case.

I wish to argue that it is more plausible to think that the
past is finite than that the quality of the restricted present
oscillates. It is true that in the case of subsystems of the cosmos
a more extensive past does not necessarily produce a richer
and more complex objective present. Many a man grows senile,
and in spite of his greater experience acts like a sorry shadow
of his youthful self. But is it plausible to think that the same
thing can happen to the cosmos as a whole? The "totalitarian
principle" of modern physics, "all that is not forbidden is com-
pulsory," applies here. What prevents the cosmos as a whole
from fulfilling the potentialities it has within itself and from
maintaining the perfection it has acquired rather than losing it?
There is nothing outside the cosmos but God, and it does not
seem plausible that God would prevent the natural dynamism
of His creature from fulfilling itself. But if it is not forbidden,
then it is obliged to fulfill itself. In that case we must be on
the way to the final perfection of the world, and the developmental process we are engaged in must have begun a finite
time ago. The fact that the present state of the cosmos is one of
finite richness and complexity leads us not only to the conclusion that there are a limited number of levels in its hier-

archical structure but also that there are a finite number of moments in its history.

From a Christian perspective one might object that the cosmos is indeed obliged to fulfill itself but that this obligation is only a moral one. The perverse will of finite creatures is capable of stultifying the natural dynamism of the cosmos. I am in partial sympathy with this viewpoint and will argue later that the present order of the cosmos has indeed been strongly determined by sin. But it is not possible that the sins of rational creatures should totally frustrate the purpose God has written into the evolutionary dynamism of the cosmos. Furthermore, Christian faith teaches us that the harm done to the order of the world by sin has been remedied by Christ.

The arguments I have given will seem lacking in force unless one shares my vision of evolution as a purposeful process. The full explanation of this purposefulness will be given only in Chapter 7. Here I can only remark briefly that I believe that all causality, and preeminently the causality involved in cosmic evolution, is ultimately the symbolizing activity of intelligent beings. Such activity is purposeful and aims at full expression of the potentiality of the symbolizing subject. An eternal cosmos which alternates between periods of development and periods of decay is ultimately meaningless and repugnant to rational desire.

NOTES

1. See HARRISON, PEEBLES.
2. This statement must be accepted with caution. The mathematics of probability describes the functioning of a model which, as the discussion below indicates, gives only a very simplified description of the real world.
3. The question whether Neo-Darwinism is a complete theory was the subject of a symposium held at the Wistar Institute in 1966. For this conclusion, see MOORHEAD, p. 102.
4. EDEN, p. 7.
5. Ibid.
6. [Gloria B. Lubkin], "Amino Acids in Both Moon and Meteorite," Physics Today 24, No. 2 (February, 1971), 17–19.
7. SCHÜTZENBERGER.
8. MOORHEAD, p. 100.
9. MOORHEAD, p. xi.
10. MOORHEAD, p. 71.

11. MOORHEAD, p. 100.
12. HARRIS, p. 232.
13. *Confessions,* Book 11, Chapter 14.
14. *Confessions,* Chapters 15–20.
15. This view of time is similar at least in some respects to that of Whitehead. But it occurred to me without any conscious dependence on Whitehead, and I am not enough of a Whitehead scholar to assess the relationship accurately.

3

A Model for Space, Time, and Matter

1. THE DIGRAPH MODEL

I SHOULD LIKE NOW TO ILLUSTRATE some of the foregoing ideas
and to facilitate the discussion of additional ones by means of a
model. I believe that the basic concepts embodied in the model
are valid, so I expect that they will prove relevant for the future
(possibly the far distant future) of science. But the model itself
is not complete or precise enough to be scientifically useful.
There is a vast difference between a valid speculative idea and
a scientific theory, and it may be that the underlying concepts
should be worked out in a different way. Meanwhile, however,
the model will serve us after the fashion of a Platonic myth.

The concept of system is one of the basic ideas of physics; it
has never succeeded in dominating all of physics, however,
because its application to the physical realities expressed in the
concepts of space and time[1] (which together with that of matter
itself are the three fundamental concepts of physics) has never
been clear. Geometry preceded physics and furnished it with
a ready-made and very elegant model for physical space and
time, a model relatively unrelated to the concept of system.
The original Euclidean form of the geometric model proved to
be overly simple in the early days of this century and was ex-

61

tended by the introduction of non-Euclidean geometry. Since then, even more complicated geometries have been used by Einstein, Wheeler, and others.[2] I do not believe, however, that any geometric model can ever be fully satisfactory because geometry does not give an adequate account of the system concept.

We have seen that the universe is a hierarchy of systems, some of the principal levels of which are given by the following sequence: organism, cell, molecule, atom, elementary particle. The question naturally suggests itself, are the entities on the last level of the sequence, the "elementary particles," themselves systems?[3] If they are, then they are composed of a finite number of entities from a yet lower level which are interacting among themselves so as to form the elementary particles. These subentities might in turn be composed of yet smaller entities, and so on. Is there a last term to the sequence and if so why?

A key consideration is that as one proceeds downward in the hierarchy from organism to elementary particle the systems and their interactions get simpler.[4] The sequence must terminate, then, when one arrives at a level where the entities and their interactions are as simple as possible. The simplest possible entity would be one which has no internal structure, either spatial or temporal. The simplest possible action would be to form a relationship with another entity of the same type. Such an entity could be adequately represented by a mathematical point, which has no internal structure. Its interactions could be represented by ordered pairs, the first element in the pair being the entity which acts and the second the entity acted upon. Such an ensemble of points and directed pairs constitutes a mathematical structure known as a "digraph" or "directed linear graph."[5] A representation of a digraph can be constructed by putting some dots on a sheet of paper and connecting them in an arbitrary way by arrows (see Figure 1a). The dots represent the "vertices" of the digraph and the arrows its "arcs." We shall consider only those digraphs in which each arc begins on one vertex and ends on another distinct from the first, since we wish the arcs to represent the action of one entity on another distinct one. Note that if an arc goes from vertex A to vertex B, there may or may not be another arc going from B to A.

A digraph is an abstract entity, and lines and points on paper are not identical with it; they merely represent it. A vertex is not merely small in size; it has no size at all. Arcs do not have length or direction. They are simply abstract relationships involving two vertices ordered by the relationship.

Another way of representing the structure of a digraph is by means of a matrix. The rows and columns of the matrix are labeled with the labels of the vertices and if a given vertex, A, sends an arc to another, B, then a "1" is placed at the intersection of the Ath row and Bth column of the matrix. Otherwise a "0" is placed there. Figure 1b shows the matrix corresponding to the diagram in Figure 1a.

A digraph does not have the ordinary metric properties of Euclidean space or other manifolds. Still, one can define a natural distance-function by allowing the distance from vertex A to vertex B, $d(A,B)$, to be equal to the number of arcs in the shortest directed path from A to B, where it is understood that a directed path from A to B goes "with the arrows." Note that $d(A,B)$ is not necessarily equal to $d(B,A)$, as illustrated in Figure 1. One can also define a distance-function, say $I(A,B)$, in terms of paths which are not directed, that is, paths which are sometimes with, sometimes against, the arrows. This

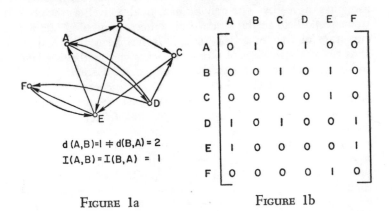

$$d(A,B)=1 \neq d(B,A) = 2$$
$$I(A,B) = I(B,A) = 1$$

FIGURE 1a   FIGURE 1b

distance-function is symmetric so that $I(A,B) = I(B,A)$. (See Figure 1.)

Neither of these distance-functions, however, provides nearly as much information about the relationship of two vertices as the ordinary Euclidean distance-function. If two points are separated by a distance of 10 in Euclidean space, one knows that the shortest path between the points is the only one which has this length and that there are an infinite number of other paths which are infinitesimally longer and infinitely close to the shortest one. In the case of the digraph, two vertices separated by a distance of 10 may have any number of distinct paths of length 10 joining them, and one knows nothing about other longer paths. There may be no other distinct paths or there may be many. The next longest path could have length 11 or length 10^5. So mere knowledge of the length of the shortest path between A and B does give much information about the relationship of the two vertices in the graph; one really must know the structure of the whole graph.

Assuming that we are right and that the hierarchy of natural systems has a lowest level which is not the level of the present "elementary particles," we ask, what is the relation of this lowest level of entities to space? Present-day physics gives no clear-cut answer to the question of the relationship of particles to space. In classical relativity theory (which is not a quantum theory and therefore incomplete), space is regarded as a physical reality. The gravitational field, as well as the electromagnetic field in the absence of sources, is nothing but the curvature of the underlying space. It has been the ideal of Einstein and a number of other physicists and mathematicians to build all particles and fields out of space.[6] If such a theory could be constructed, regions of the universe containing fields and particles would simply be regions in which space assumes a more complicated structure than it does in regions where there are no fields or particles. So far, however, no one has succeeded in the quest for such a theory.

At the present time, standard quantum theory still presupposes geometrical space which is then filled with fields and their particles. This geometrical space, however, seems to be nothing but a mathematical peg to which fields may be attached.

Model for Space, Time, Matter   65

Space is never simply empty. As early as 1930, Dirac's theory of the position suggested that space is filled with an infinite sea of electrons occupying all energy levels up to a certain value of the energy, the "Fermi level." The electrons which we observe are those which have acquired extra energy and risen above the Fermi level, leaving behind a "hole," which is observed as a positively charged twin of the electron. Developments of the past forty years have shown that the collection of particles below the Fermi level, the "Fermi sea," is not inert but is in a state of ceaseless agitation in which pairs of particles are continually emerging from and being reabsorbed into the "vacuum." The values of charge and mass for all observed particles are strongly influenced by their interactions with the vacuum.

The prevalence of violent, particle-like excitations of the vacuum in modern physics has led to a suggestion which is in some respects the opposite of the "space-is-all" viewpoint stemming from relativity. Chew has suggested that particles are fundamental and that space should be constructed out of particles in relationship to one another through interaction.[7] This view seems to me to embody a fundamental truth. But the "particles" from which space should be constructed are not the elementary particles of contemporary physics. Both the vacuum and the elementary particles are subsystems of the universe and are composed of elementary entities from the lowest level in the hierarchy of natural systems. It is for this reason that elementary particles can emerge from and be reabsorbed into the vacuum. The entire physical universe should be represented by a single digraph.

The idea that space should be represented by a digraph can be arrived at from a slightly different, though related, point of view. The usual model for physical space is a "manifold" in which each finite volume contains an uncountably infinite number of points. The precise logical status of the infinite and of continua, which are conceived of as the union of an infinite number of points, is a difficult intellectual problem which has exercised generations of mathematicians even to the present day. In a famous lecture delivered in 1925, the mathematician David Hilbert insisted that "the infinite" can only be legitimately regarded as an "ideal element" and that we are not justified in ascribing any kind of reality to the "actual infinite." [8] Whatever may be the final status of the various types of infinity in mathematics, the use of continua as *physical* models compounds the problem by implying the existence of an infinite number of *physically* real points in a finite volume. As Hilbert insisted, it seems that actual infinities are never found in physics and that physical quantities are always finite and discrete. The "actual infinite" finds its way into physics only through geometrical models for space–time. The readings of scientific instruments are finite numbers, not infinitesimals. Therefore we need not think the infinitesimal structure of the manifolds used by physicists today corresponds to anything real. It has been pointed out that a clever mathematician presented with a sufficiently detailed catalogue of airline distances between various points on earth could arrive at our usual geometrical sphere model as the simplest way of summarizing all that data. We shall assume that manifolds are merely the simplest way of summarizing current information about the paths between vertices of a digraph, vertices which are always separated by finite distances.

As long as space was merely the inert stage on which the real actors, particles and radiation, played their role, no questions arose about the utility of the continuum model. But during the present century space itself has been getting actively involved in the dynamics of physical theories, and during the past twenty-five years inconsistencies have been appearing in the most fundamental theory of physics, quantum field theory, inconsistencies which seem to be associated with the assumption that there are an infinite number of points in every neighborhood of space. A number of physicists have suggested that the remedy is to assume that space is discrete, that is, that a finite volume contains only a finite number of points. Once this idea is accepted, it is natural to think of such points as physically real and related to one another through interaction. Then the digraph model suggests itself as a way of representing such a system.

Both modern physics and this book treat the material universe as a four-dimensional structure. There is a difference inasmuch as my model has a beginning and terminates at the present moment while the model of physics simply extends indefinitely into past and future. Both models, however, commit one to accept at least partial homogeneity between space and time. Consequently, in the digraph model time as well as space is discrete. A finite interval of time, indeed all time to the present moment, consists of a finite number of moments which can be labeled by the nonnegative integers. We divide the entire set of vertices which represent all elementary entities into sets labeled with the same integers. If an elementary entity comes into existence at time T, then its representative vertex belongs to set T. We shall assume that all arcs beginning on set T terminate on set $T + 1$.

The universe and the systems in it seem to be characterized by conservation laws, so one is inclined to ask, what is conserved in this digraph model of the universe? The most obvious guess would be that the number of arcs leaving a vertex is the same as the number entering it. Since arcs represent action, this rule corresponds to the classical philosophical axiom that an entity can act only to the extent that it has previously been actuated. Clearly the rule does not apply to the zeroth vertex-set or to the one corresponding to the objective present. Since the zeroth set is first, it receives no arcs. Since the set corresponding to the objective present is the last, it sends no arcs. However, at the next moment of time it will send to the new objective present the number of arcs which it is now receiving.

If a digraph model of the universe is to be correct, it must have a structure which on the scale accessible to our present-day instruments averages out to give the properties to which we are accustomed. One can draw an analogy by considering the relationship between the fluid model and the atomic model of a gas. The instruments and techniques of the eighteenth and early nineteenth centuries furnished physicists with data which were most easily interpreted by thinking of a gas as a rarefied, highly compressible fluid. In our day we know this model to be an approximate one, valid for a certain domain, and believe that the physical structure is better understood by considering the gas to be a collection of atoms in rapid motion, colliding frequently with one another.

Similarly, physicists today regard the classical concept of the path of a particle in space as a way of expressing the average or expectation value of behavior which is more complex.

The data about the cosmos available to physicists today are most easily understood by thinking about it in terms of geometrical manifolds, fields, and particles. But philosophical considerations persuade me that the concepts of entity and action are more ultimate than space and time, and that conviction, together with the observed hierarchical structure of matter, leads me to postulate that the digraph model will ultimately give a better approximation to physical reality than the current ones. The average properties of large subgraphs of the total digraph will be the same as those of present-day models, just as large numbers of atoms of a gas have collective properties equivalent to those of a rarefied compressible fluid, or the quantum mechanical behavior of an ensemble of electrons averages out to the classical trajectory.

We may, perhaps, visualize this correspondence in the following way: Since the digraph model is associated with a universal cosmic time, it is convenient to define a set of "spatial graphs," each of which corresponds to a single moment of time. The spatial graph for time T consists of a set of vertices, one for each elementary entity which is in the objective present at T, and a set of "edges." The set of vertices can be identified with the vertex set T from the primary digraph model. An edge is simply an unordered pair of vertices. Vertices A and B in spatial graph T are connected by an edge, $(A,B) = (B,A)$, if and only if there is some vertex C in vertex set $T - 1$ of the fundamental digraph model which sends arcs to A and B. In other words, if an elementary entity acts upon two others, then their representative vertices are connected by an edge in the spatial graph which contains them. A spatial graph may be represented by dots on a sheet of paper connected in an arbitrary way by lines, as in Figure 2a. It may also be represented by a matrix whose rows and columns are labeled with the labels of the vertices and which contains a 1 at the intersection of the Ath row and Bth column if and only if A and B are connected by an edge, and zeros elsewhere. This matrix is symmetric (that is, the values in the positions A,B and B,A are the same) as is illustrated in Figure 2b.

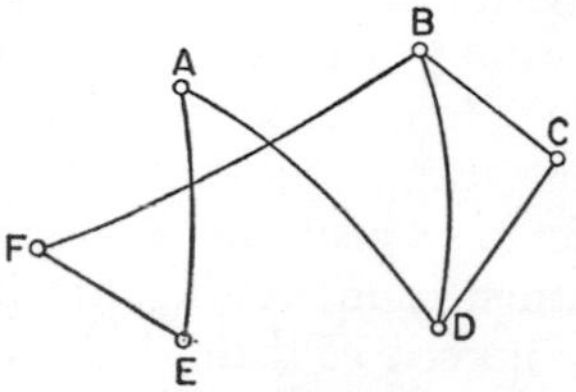

FIGURE 2a

	A	B	C	D	E	F
A	0	0	0	1	1	0
B	0	0	1	1	0	1
C	0	1	0	1	0	0
D	1	1	1	0	0	0
E	1	0	0	0	0	1
F	0	1	0	0	1	0

FIGURE 2b

We now embed the spatial graph under consideration in a higher dimensional Euclidean space in such a way that its edges are straight lines of length 2 in this space. The value 2 is chosen so that the length of the edge between two vertices in a spatial graph corresponds to the length of the path between them in the primary digraph model. We suppose that this requirement, possibly in conjunction with some other simple requirements, will turn the spatial graph into a rigid structure in the embedding space, after the manner of a geodesic dome in ordinary three-dimensional space. If one takes straight steel members of the same length and connects them in accord with the proper connection-matrix, then the resulting structure is rigid and approximates a hemisphere. By using shorter pieces, the approximation to a spherical surface can be improved. If one wishes, other geometrical shapes can be approximated by using different connection-matrices. The embedded spatial graph ought to approximate a three-dimensional curved manifold since we know from present-day physics that such a manifold gives a good representation of the average characteristics of a spatial cross-section of the universe.

Regions containing particles cannot be represented so simply and require numerous extra symmetries and additional quantum numbers. Physicists who wish to explain everything in terms of geometry regard the regions which contain particles as places in which the topology of space is more complicated than elsewhere. Similarly, we would have to say that the complexity of spatial graphs varies and that more complicated subgraphs appear to us as particles, less complicated ones as empty space. Subgraphs representing particles would not fall along the curved three-dimensional manifold which represents the average characteristics of the universe as smoothly as subgraphs representing empty space. However, differences may not be so great as we might think. Particles are extremely interesting and important, especially to us who are composed of them, but from the viewpoint of inanimate matter, they may well be like clouds in the sky, conspicuous but far less substantial in regard to the energies involved than the background of the vacuum from which they arise.[9]

Newton once wrote that "action at a distance" is so absurd a notion that no philosophically competent person could ever entertain it. Many others have shared this opinion. The insight which seems to underlie it is that for one finite entity to act on another some previous relationship must exist. Our powers of action are not so unlimited that they can be effective of themselves alone. Certain conditions must be fulfilled; the agent and what it acts upon must be related in a particular way. Experience seems to indicate that the previous relationship required involves spatial contact. Without contact we can exert no force, at least as far as we are able to determine on the basis of naïve experience and of contemporary science alike.

In the digraph model, however, the relation between nearness and action is more complex. The spatial relationship of one vertex in the objective present to another depends upon the number of arcs in the various paths between them. One might say therefore that nearness is determined by action rather than the reverse. This statement, however, needs to be completed by pointing out the fact that the entire structure of arcs and vertices which constitutes the objective present is determined by the structure of the entire past. (This will be discussed at greater length in section three.) If one is speaking of the complex patterns of arcs and vertices which comprise macroscopic entities, one would have to say that the relative location of particles in the past determines the way in which they relate to one another in the objective present. But relative location in the past is very much a function of action in the past.

In the final analysis, therefore, action is more fundamental than proximity and contact. This is philosophically more satisfactory than the usual view. I do not regard the concepts of space, time, distance, etc., as fundamental. They are precipitates of our sensible experience, strongly determined by the type of data processing which occurs in the genetically determined structures of our vegetative and animal levels. Ultimate concepts are "metaphysical" ones which derive from our human experience as intelligent and purposeful beings. Such are "entity" and "action," which are represented in the digraph model by vertex and arc. It is more satisfying to derive the notions of distance, nearness, contact, etc., from the structure of vertices and arcs than to postulate them as fundamental. This is but a particular case of the general thesis of this book that concepts derived from the higher levels of being are the fundamental ones in terms of which the concepts germane to lower levels must be understood. When we come to understand the cosmos more fully, we shall see that even the concepts of physics are derived from those concepts which characterize intelligence and freedom.

2. RELATION OF THE MODEL TO RELATIVITY THEORY

It is clear that there is some problem about reconciling the structure of our model with relativity theory. The usual interpretation of the special theory of relativity asserts that given a particular physical event, there is no unique way of singling out a set of events which are simultaneous with it. Nature knows nothing of absolute simultaneity; temporal relations between spatially separated events are (within certain limits determined by the speed of light) fixed only by an arbitrary choice of reference system. In the digraph model, however, each elementary entity is a member of a single, uniquely determined set of entities which come into existence at the same moment and retain their interrelationships even when they have become part of the past. The event which is the creation of one of these entities is absolutely simultaneous with the creation of the other entities represented by the same vertex set. Therefore there is a single universal time for the whole universe, each of whose moments is defined by the emergence of a particular set of entities.

This picture is not so alien to general relativity as it is to special relativity. Relativistic models of the expanding universe often assume a unique common time for the universe. It is chosen so that the space corresponding to a given instant of time is one in which the total momentum of the mass and the radiation in the universe is zero. But in the digraph model the elementary entities are divided into simultaneous sets in a much stronger way. One should expect that this strong classification would correspond to a strong physical cause which would be a potent source of physically detectable effects, whereas in reality symmetry with respect to choice of a time axis is one of the most obvious features of present-day physics. One must recall, however, that the parallax of the stars was not discovered until the nineteenth century in spite of the fact that it is an obvious consequence of heliocentrism. Nowadays, this does not seem surprising to us because we know the very good physical reasons why the effect is so small and so difficult to detect.

In the early days of this century, Einstein took the absence of physical evidence for absolute simultaneity as a sign that its possible existence could be disregarded, and he concentrated on developing the consequences of the resulting symmetry in physical law. His approach proved to be enormously fruitful. Our modern experience with partial symmetries, however, should predispose us to wonder whether Lorentz symmetry (i.e., symmetry with regard to the choice of temporal relationships) may not also be partial, even in microscopic problems where the global structure of the universe is not considered.[10]

A strong argument for this is one connected with the nature of consciousness. We are conscious of ourselves existing as a unitary whole *now*. The "now" embraces our being in a simultaneous whole. Consequently, all the physical events which are part of the structure of the now existing self are simultaneous in a physically unique way. If one holds a Cartesian philosophical position in which the self is a purely "spiritual" being making contact in an extrinsic way with matter at a single point in space, then this fact presents no problem. The "now" of the human spirit is simultaneous with the "now" of the single point in space at which the spirit interacts with matter, and there is no reason to demand that it have a unique temporal relationship with events at any other point in space. However, if one holds that the reality of a human self includes matter, then it is difficult to see how one can deny absolute simultaneity between the events in the human body which are integrated into the conscious experience I am *now* having.

The human self is a system which includes, at the very least, many of the neurons of the brain. The reality of the whole is dependent upon the properties of the parts it integrates into unity, and therefore the present state of the whole is a function of the present state of the parts. Since the state of the whole is a uniquely determinate reality *now,* so are the states of the parts which contribute to it. We thus have a finite volume of space, that of the brain at least, in which a unique present can be defined.[11]

It was one of Einstein's great insights that simultaneity can only be defined in terms of physical interactions. The reason for this is that simultaneity is a relation which is actually established by interactions. Therefore the fact that different events in the brain are really simultaneous means that the parts of the brain which contribute to consciousness are interacting immediately. More precisely, these various subsystems are interacting immediately with the whole since they are simultaneous because of their contribution to the present state of the whole. Lorentz symmetry is broken, therefore, by the physical interactions which characterize consciousness. It is clear, as will be discussed at length later, that some of these interactions are of higher order than any known to contemporary physical theory. Hence there is no need at present for physicists to concern themselves directly with such interactions. But the existence of these interactions means that Lorentz symmetry is not a fundamental constraint on all physical processes, and thus it is possible that there may be interactions within the domain of present-day physics which also break it.[12]

3. LOCAL AND HOLISTIC ASPECTS OF DEVELOPMENT

Assuming then that there is indeed a unique time for the whole universe, we consider again the issue of the general law which determines the development underlying this time. What is the prescription for deriving the objective present from the past? In terms of the spatial-graph model, what is the rule for deriving graph T from the set of graphs $t < T$? We have already stipulated some simple requirements for the digraph model, and corresponding conditions will obtain for the spatial graphs derived from it. But beyond this to say anything both specific and plausible is an extremely difficult task. Instead let us consider the very general and philosophical question of whether the complete law of development of the digraph is local or holistic.

Contemporary physical theories are local in character. This is to say that an event occurring at point P at time T is determined completely by events within the "backward light cone" [13] of P which are infinitely close to P both temporally and spatially. Events far removed from P in both time and space do indeed affect it, but their influence is mediated by a long causal chain whose last links are the events in the infinitesimal neighborhood of P. Thus we do not have to know what is happening in the Andromeda galaxy in order to predict a train of events here on earth. If the Andromeda galaxy suffers a catastrophe which affects us here, we will know of it only long after the event in terms of its transmitted radiation whose arrival is an event occurring here on earth. Similarly, it does not matter whether a satellite was put into orbit ten microseconds or ten centuries ago as long as its present position and momentum are known. This local character of physical law is the reason we have postulated that arcs arriving at vertex set T must originate in set $T - 1$. In the digraph model, the equivalent of the backward light cone of P is the set of vertices from which there is a directed path to P. If Q sends an arc to P, then Q is within the equivalent of P's backward light cone and is also as close to P as possible.

Although physical laws are local, there is another aspect to be considered. What is the status of these local laws themselves?

Are they disembodied forms hanging in some Platonic heaven? Or are they rather formal intelligibilities which reflect the nature of the fundamental dynamism of the cosmos? We have already opted for the latter alternative. Note that it is the dynamism of the cosmos *as a whole* which grounds physical laws.[14] That this is true is clear from the fact that the same laws govern all the matter in the universe regardless of the distance separating its particles from immediate interaction. Consequently there is a holistic aspect to the causality by which the objective present derives from the past.

We are dealing with two different kinds of causality, the local causality which is represented by an arc which links one vertex with another on which it acts, and a more indefinite, hazy kind of causality by which each part influences the whole and the whole each of its parts. This seems to be akin to the distinction which Aristotle conceptualized in terms of "efficient" and "formal" causality. Efficient causality is the kind of causality I exert when I hit a typewriter key or do any kind of physical work. Formal causality is the kind of causality an idea exercises on the set of words which express it. The subject of a sentence does not cause the verb to agree with it in regard to number by means of efficient causality. In a sense it does not exert causality at all. Yet the fact that the subject is plural entails certain consequences for the form of the verb. Similarly, the state of each part of the material universe entails consequences for the other parts. It is plausible to suppose that the framework which contemporary physics takes for granted, i.e., the three-plus-one-dimensional space–time, as well as the empirical constants, interactions, physical observables, symmetries and quantum numbers, and finally the local laws themselves, are reflections of the state of the whole universe with all its parts.

We shall conclude ultimately that matter is the symbolic self-expression of a thinker or set of thinkers and that its fundamental dynamism is a knowing desire or desireful knowing. It is because of the nature of thought and symbolizing activity, which have to do with the formation of a coherent *Gestalt*, that the characteristics of each part of the cosmos entail consequences for every other part and for the whole.[15] Just as a man perceives the details of the situation which confronts him in relation to his own intentions, and then produces a decision which alters the initial situation, so the dynamism of cosmic evolution is influenced by the configuration of each part of the cosmos and in turn forms it into a new and coherent *Gestalt*. In accord with the assertions of the preceding chapters, this process has two aspects, a mechanistic and a creative one. The workaday business of the cosmos is controlled by mechanisms, that is, by "inviolate laws of nature," but these mechanisms are subject to the organismic regulation of the creative dynamism which has established them to fulfill its purposes.

4. EXTENSION OF THE DIGRAPH MODEL TO HIGHER-LEVEL SYSTEMS

The digraph model is primarily intended to facilitate the discussion of concepts such as space, time, simultaneity, causality, etc., which are germane to physics. We can extend it to take into account higher-level entities as well. But the extended model is probably a much less adequate representation of the higher-level systems than the original one is of the lowest level. One cannot properly represent different levels of existence on the same diagram.

Higher-level systems do not exist apart from the elementary entities of the lowest level. They organize them into unities more complex than the powers operative on the lowest level themselves could achieve. Consequently a higher-level system will be a region of the universe which exhibits greater complexity in virtue of an organizing element which is not reducible to elementary entities. We therefore introduce into the digraph "vertices of the second kind." Each such vertex is an organizing element for a set of "vertices of the first kind" which represent the elementary entities included in a higher-order system. The vertices of the first kind in the set exchange arcs with the vertex of the second kind, which functions as a switching center rerouting arcs among the vertices of the first kind, thus creating a pattern of action which could not be had without it. We shall assume that a vertex of the second kind is included in the same vertex set as the vertices of the first kind with which it interacts. The arcs which it exchanges with them therefore begin and end in the same vertex set and represent a type of causality which does not proceed from past to objective present. These arcs, which we can call "arcs of the second kind," represent formal rather than efficient causality. Nevertheless we shall still assume that arcs are conserved as before. Figure 3 illustrates the way in which a vertex of the second kind might modify the pattern of action of a set of vertices of the first kind.

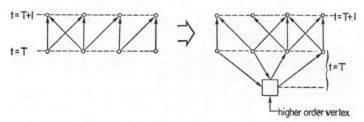

FIGURE 3

Higher-level systems like man have not only a complex spatial structure but a temporal structure as well. That is to say that men perdure through time in a way unlike that of the elementary entities which come into existence with an already complete structure. Ours continues to develop. We shall assume therefore that vertices of the second kind occur in strings which run through a number of vertex sets. Each vertex of the second kind is really the intersection of a "world line" with a particular vertex set, and the successive vertices of the second kind along the same world line are identified with one another in a way in which vertices of the first kind connected by arcs are not.

Vertices of the second kind can exchange arcs of the second kind with one another, thereby uniting themselves and their subordinate vertices of the first kind into a system which has no one center of unity. They can also give rise to a higher, third-level, entity which is represented by a "vertex of the third kind." A vertex of the third kind does not exchange arcs with vertices of the first kind; it serves, rather, as a switching center rerouting arcs among vertices of the second kind and changing the normal pattern of action which would prevail if they were left to themselves.

In general we introduce a new kind of higher-order vertex for every irreducibly different level. A higher animal such as a dog or a chimpanzee would therefore contain at least four different kinds of vertices. Besides the necessary vertices of the first kind which represent the elementary entities, it would have to contain a second kind to represent its vegetative systems, a third kind to represent its sentient systems, and a single vertex of the highest kind to represent the final unity of the animal system, which corresponds to the conscious self in man. It is quite conceivable, of course, that there may be many more than four irreducibly different levels within an animal and, if so, there would be more than four kinds of vertices in its digraph representation.

It is not evident that lower animals and plants have a single supreme center of unity. If they do not, they would have a digraph representation in which there are many highest-level vertices interacting among themselves to form a system without a single center. Such a system would be more complex than any one of its parts, but it would not be on an irreducibly higher level than the parts. The total system would be a "higher-order machine" since, even though the parts would be vegetative or sentient, they would not be changed by insertion into the whole but merely coordinated on the same level by vegetative or sentient interactions, much as the parts of an ordinary machine are coordinated on their own level. In a similar way, on a level lower than that of life, a crystal is not irreducibly different from one of its unit cells.

We see that "mechanism" and "organismic regulation" can occur on any level and in various combinations. Mechanism implies coordination on the same level as that of the parts by interactions characteristic of that level or of a lower level. Organismic regulation implies coordination of the parts by a higher-level system into which they are taken up. The interactions involved here are of a higher type than that which the parts are capable of by themselves, and they bring the parts into immediate, simultaneous contact with the whole.

In man there is a single conscious center, so organismic regulation is the ultimate coordinating factor. Since this regulation does not touch all the vegetative and sentient systems of the body directly, not all parts of the body belong to the person in the same sense. Only those parts which contribute to the simultaneous whole given in consciousness are constitutive of the personal whole. The fact that it takes time for an injury to my foot to affect my consciousness indicates that interaction between my foot and me is not totally on the personal level. Many organs have a life which is essentially the same as that of a lower animal or plant and are linked to the activity of the personal center by interactions below the personal level. They are certainly mine and a part of me but not in as full a sense as those portions of the nervous system whose activity contributes directly to consciousness. The degree of unity between the person and his various bodily parts ranges all the way from that had by hair or fingernails to that of the essentially human structures of the nervous system. This reflection helps to eliminate various conundrums posed for the holistic thinker by such things as organ transplants, blood transfusions, etc.

Perhaps it will be well to point out explicitly that no one will ever succeed in isolating a higher-order vertex as physicists have isolated protons, neutrons, and electrons. When one takes apart a living cell, what remains is a set of molecules. The higher-order vertex (or vertices) which gives it unity disappears with that unity. Similarly, no one will ever succeed in measuring the length of a path through higher-order vertices by the methods of physics. Lower-order interactions cannot traverse such paths.

It seems likely that the number of arcs incident with a higher-order vertex is much smaller than the number which comprises the fabric of its subordinate systems. In other words, the behavior of subordinate entities is much the same as it was before their integration into the superordinate system. It is by means of this relatively small number of arcs, however, that the mechanisms needed for life are constructed and regulated.

The extension of the digraph model which we have made has some advantages since it assists in making our ideas about higher-order systems more concrete. But the extended model is probably a much less adequate representation of these higher-order systems than the original model was of the lowest level. The higher-level entities represented by higher-level vertices are not adequately distinct from their subordinate entities. Furthermore, the action of higher-level entities is essentially different from that of lower-level ones and is not simple. Hence representing it by arcs, even if we label them differently by calling them "higher-order arcs," is much less satisfactory. The activity of a second-order entity consists in the organization of the pattern of activity of the first-order entities (which, we must remember, are not adequately distinct from it). The activity of a third-order entity affects the organizing activity of second-order entities. It is an organizing activity which organizes an organizing activity. No digraph can adequately represent this.

A correct understanding of the whole universe must make use not only of quantitative concepts but also of analogous ones which are not susceptible of total explicitation in simpler terms. The relationship of the various members of the natural hierarchy to one another is perhaps best modeled on the relationship of a complex notion to the notions used in defining it or of the intelligibility of a sentence to the intelligibility of the words which compose it. A noun or verb does not lose its own meaning by entering into a sentence; rather that meaning is taken up and modified by the sentence into which it enters in a way which cannot be completely formalized but understood only implicitly in understanding the sentence. The reason why these models are relevant is the fact that the universe is ultimately the expression of thought and cannot be fully understood in terms of anything lower on the ontological scale.

Nevertheless, if we are aware of its limitations we can think in terms of the digraph model. Thus we can speculate about the possibility that the higher-order vertices which represent human persons may at times exchange arcs directly rather than through lower-order structures representing sensory-motor faculties. If there is such a thing as extra-sensory perception, this would provide a model for it. We can also begin to consider the possibility that these vertices of the human level also exchange arcs with a single highest-order vertex which represents the unified being of the entire universe, the Teilhardian "Omega point."

Clearly enough, the comprehensive law describing the time

development of the entire universe will have to be such as to take account of entities of higher order. In fact if their appearance during the evolution of the universe is not to be supernatural, this law will have to be responsible for their appearance. A living system, and the higher-order unities which characterize it, will have to appear spontaneously whenever a subsystem of the universe has developed a sufficiently rich structure. It is clear then that this comprehensive law does not follow rigorously from any number of postulates which deal with graphs not containing higher-order vertices. In other words, if physics and chemistry are sciences which establish their laws by doing experiments on structures below the level of life, then they will never find the total law which governs not only inanimate matter but living systems as well.

In reality this is nothing new to anyone not wearing reductionistic blinders. Even in physics the law governing the interactions of charged particles is obtained by studying systems of charged particles. The law governing the interaction of nucleons is obtained by studying systems of nucleons. No one has ever claimed to be able to get the form of interactions by studying free particles. Why then should one expect to be able to deduce the form of interactions in living systems from the behavior of systems below that level? We know that we have a new type of system only when we have observed a new type of interaction.

NOTES

1. It is true that spatial and temporal position are involved as parameters in many systems. When I suggest applying the concept of system to space and time, however, I mean something much stronger, as will soon become evident.

2. See WHEELER and MISNER.

3. This is an important question in present-day physics. See WEISSKOPF and CHEW (II).

4. I consider this statement valid. However, one could get into interminable discussions trying to specify the precise meaning of "simpler." I prefer to affirm only that it is evident that a human society is more complex than a molecule and that the same kind of qualitative simplification occurs with each downward step in the hierarchy. In the end one will probably judge the correctness of this heuristic assumption in terms of the coherence of the picture which results from making it.

5. HARARY provides a good introduction to graph theory for those
with a certain amount of "mathematical maturity."
6. See WHEELER and MISNER. Note that in the present context a
"space" means a "space–time."
7. CHEW (I), (II).
8. See HILBERT.
9. I believe that this analogy, like several others I have used, is due
to J. A. Wheeler who, in addition to his fundamental scientific work,
has written excellent popular summaries of relativity theory.
10. A physical system possesses a symmetry when it can exist in
several different states which are indistinguishable as far as a set of
relevant physical observables is concerned. Thus one can rotate a
perfectly symmetrical top about its axis of symmetry, and its state at
the end of the rotation cannot be distinguished from its state at the
beginning. If the top is perfectly symmetrical except for the fact that it
has a mark painted on it at a certain point, then the symmetry is
partial. A color-blind person who could not see the mark would think
the top perfectly symmetrical, but if his color blindness were corrected,
he would realize that the top is only partially symmetrical. At present,
physicists cannot see that the choice of a particular time axis, which is
to say, the choice of a particular set of events as simultaneous, makes
any difference in the way a physical situation develops. They therefore suppose that the symmetry associated with this choice (Lorentz
symmetry) is total. However, the progress of physics may bring to
light some interaction which depends upon the choice of time axis and,
if this happens, Lorentz symmetry will be revealed as a partial one.
11. The fact that the time required for a neuron to fire is much
longer than that required for light to traverse distances such as those
within the human body is essentially irrelevant. Let us think of the
state of the brain as defined by listing which neurons are "on" and
which are "off." Then if a certain neuron switches its state, the state
of the brain, and presumably of consciousness, ought to change at that
very moment. As a result the temporal order of neuron switchings in
different parts of the brain is determinate and not indeterminate as
special relativity would have it. More complicated definitions of the state
of the brain do not change the essential point of this argument.
12. The possible existence of a physical phenomenon which would
do this is discussed in BILANIUK.
13. The backward light cone of point P is the set of points from
which a signal can be sent in time to arrive at P. This is the same
as saying that the set of points in space–time at which an event can
occur will produce effects at P.
14. SCIAMA contains an interesting discussion of the way in which
distant matter might produce the property of inertia in matter here on
earth. This idea, due to Berkeley and Mach, has been developed by
BRANS and Dicke into a full-fledged theory, which may possibly be a
superior alternative to Einstein's gravitational theory.
15. That this is the case is the main thesis of SCIAMA, though he is
far from attributing its truth to the nature of consciousness.

4

Personal Knowledge

1. PIAGET'S THEORY OF INTELLECTUAL DEVELOPMENT

ULTIMATELY THE COSMOS can be understood only as an expression of thought. In this chapter we begin laying the groundwork for such an understanding by considering the ideas of Piaget, Kuhn, Koestler, and Polanyi about the nature of human intelligence and its creative acts, especially in the field of science.

The Swiss psychologist Jean Piaget has been studying the development of human intelligence for more than forty years. The result of his labors and of those of his associates is a remarkable body of psychological theory and experimental data which is still growing. I make no attempt even to sketch all this material.[1] I shall merely single out some of Piaget's fundamental concepts which I believe contribute significantly to a general understanding of the cosmos. Since these concepts have been rooted in experiment by Piaget and his followers, they help to give some stability to the highly theoretical and speculative structure which I am erecting.

Piaget began his career as a biologist, and his notion of human intelligence has a biological cast. For him the act of understanding is an action of the human organism by which it relates and adapts itself to the world. Adaptation has two aspects, assimilation and accommodation. By assimilation the person appropriates the world to himself, first in a purely physical way, later in a symbolic way. By accommodation he changes, enlarges, and perfects his own intrinsic structures. The goal of adaptation is to attain a state of "equilibrium." This word has unfortunate static connotations. But Piaget intends by it a dynamic state of relationship with the environment in which the powers of the organisms are engaged to the full in activity which is rewarding for its own sake as well as useful in sustaining life.

For Piaget, intellectual structures are not inherited but constructed. What is inherited is an initial level of organization which is not intellectual, and a mode of functioning which makes possible the development of further, and eventually intellectual, structures. Each action or functioning of the organism grows out of an existing structure and terminates in an enlarged, more adapted and mature, structure. Thus organization and adaptation are "functional invariants" which are found at every stage of development and which characterize the very nature of the intellect in a way in which particular structures or contents do not. These functional invariants characterize not only intelligence but biological functioning in general.[2] Intelligence is therefore a solidly biological phenomenon, deeply rooted in the nature of the universe. The development of intelligence in each human person is the continuation of the general evolutionary development of the cosmos.

It can be seen that for Piaget genetic information is less like a blueprint than an algorithm or set of directions: "Stay on Route 35 till you reach the third traffic light; then turn left. Your goal is the third house on the right." This type of information is far less extensive than that contained in a map, but in many circumstances it is more effective. It depends upon the existence of a suitable environment with which the directions are correlated. Similarly, man and the world form a coherent system, and it is only within that system that our inherited mode of functioning can produce intellectual structures.

Piaget is convinced that the development of intellectual structures occurs in several qualitatively different periods. These are: (1) the period of sensorimotor intelligence (0–2 years); (2) the period of preparation for, and organization of, concrete operations (2–11); (3) the period of formal operations (11–15). There are a number of stages and substages within each of these periods. During the sensorimotor period, the infant moves from a completely reflex type of functioning to a "relatively coherent organization of sensorimotor actions vis-à-vis his immediate environment. The organization is an entirely 'practical' one, however, in the sense that it involves simple perceptual and motor adjustments to things rather than symbolic manipulations of them." [3] The concrete-operational period begins with the first crude symbolization and concludes with the onset of formal thought in early adolescence. By this time the child possesses a set of well-organized cognitive structures which enable him to deal symbolically with the world. Yet these structures, with their component symbols and operations, are for the most part bound to the concrete world and are known only implicitly as they are used to deal with particular situations. In the period of formal operations these limitations are removed. The operations of the previous period are further elaborated and become the object of reflection so that a system of operations on operations develops. The concrete-operational child had to master one by one the concepts of conservation of mass, weight, and volume, which for him were unrelated. The formal-operational adolescent can grasp the abstract idea of conservation, of which all particular conservation laws are examples. In general he becomes capable of fitting the real, concrete world into a realm of possibilities provided by his own intellectual structures.

These stages are hierarchically related so that the operations of the later periods organize and make use of the already established structures of the earlier periods. Thus the sensorimotor "schemas" become symbols during the concrete-operational period, and the symbolic operations of the latter are in turn manipulated and operated upon by the formal operations of the adolescent. Furthermore, the operations of each stage are not a mere collection of different acts, but form a highly structured whole so that the function and usefulness of one operation can only be fully understood in relation to the other operations of the same period.

Piaget likes to express in mathematical terms his understanding of the structures he discovers. He believes that the intellectual operations of the concrete-operational period can be modeled by a set of logico-mathematical structures called "groupings." Similarly, he has concluded that the core system of the formal-operational period is "an integrated lattice-group structure, not just partial and incomplete lattice and group *properties*, as in the concrete-operational groupings, but a full and complete lattice and a full and complete group, both integrated within the one total system." He also "has attempted to specify certain substructures which derive from the general *structure d'ensemble* just mentioned. These substructures, called *formal operational schemas*, are more limited and specialized cognitive instrumentalities which rotate to the fore when certain kinds of problems confront the subject." [4]

Piaget himself has been concerned mainly with the development of intellectual rather than of affective structures. He has expressed the view, however, that affectivity and intellectual activity are inseparable aspects of all conduct. This view is supported by his observation that intellectual structures tend to exercise themselves. "In Piaget's expressive phraseology, the organism simply has to 'nourish' his cognitive schemas by repeatedly incorporating reality 'aliments' to them, incorporating the environmental 'nutriments' which sustain them. As Piaget repeatedly states, assimilation is the dominant component of intelligence. And the principal attribute of assimilation is *repetition*—the intrinsic tendency to reach out into the environment again and again and incorporate what it can." [5] Thus, just as the existence of bodily structures capable of assimilating food is necessarily accompanied by a desire to eat, so the existence of intellectual structures is necessarily accompanied by a desire to understand.

In fact, this desire is an aspect of the very act of existing of the intellectual being, as I shall argue later.[6] This viewpoint clearly implies that there ought to be a strong correlation between affective and intellectual development. The existence of such a correlation has been verified by Décarie.[7]

The work of Piaget and his associates provides an empirically grounded extension of insight into the hierarchical structure of the universe. The same basic pattern which we have discussed at length above reappears on the specifically human level. Once again we find subordinate systems being united among themselves to form a superordinate system which depends upon and uses their properties and, at the same time, transforms them and enables them to contribute to higher-order activities of greater ontological depth. The development of intelligence can therefore be regarded as the continuation of evolution. The unity of the whole process is often expressed today by saying that biological evolution is a process of "genetic learning," the results of which are coded in the genes, and that individual development continues the learning process on another level, coding the results in the brain and nervous system.

2. KUHN'S THEORY OF THE DEVELOPMENT OF SCIENCE

Teilhard de Chardin saw not only individual intellectual development but the cultural development of the whole human race as the continuation of evolution in the human sphere. These two kinds of development are inseparably united. Man is a social being, and the individual cannot develop except in a cultural context. Conversely, cultural development is impossible unless it is carried by the personal development of individuals. One of the most important chapters of modern cultural development has been the rise of science. In an influential book, *The Structure of Scientific Revolutions,* Thomas Kuhn has pointed out some characteristic features of the progress of science which fit in well with the ideas of Piaget and contribute to my own inquiry. To begin, I shall discuss Kuhn's ideas in the form which they assumed in the first edition of his book. Then I shall mention the qualifications he has made in the second edition.

Kuhn emphasizes the importance of what he calls "paradigms." These are successful pieces of scientific work which serve as models for the subsequent efforts of scientists. The acceptance of a paradigm, such as Newton's work on the solar system, involves the adoption of concepts, methods, attitudes, and interests which are embodied in the paradigm, which provide a framework and guide for subsequent efforts. Within such a framework, what Kuhn calls "normal science" becomes possible. Normal science is a persevering, cooperative effort by the community of those who accept the paradigm to account for as much as possible of the universe in terms of the basic framework contained within the paradigm. This effort is at once perspicacious and blind. It is blind, at least to a certain extent, because problems and evidence which cannot be handled within the accepted framework are often not seen or are regarded as uninteresting and "unscientific." But at the same time it is perspicacious because the paradigm-based framework enables even ordinarily intelligent men to grapple with certain problems in a very tenacious, persevering, minutely precise, and penetrating way. Furthermore, the fact that everyone in the particular scientific community accepts the same set of paradigms means that communication is relatively quick, easy, and reliable. As a result the potentialities of the basic framework are developed to the fullest possible extent, and the set of problems which can be handled within it are solved comparatively expeditiously.

But there are always some problems which, from the very beginning, resist solution, even problems which the paradigm-based methods of normal science seem capable of handling. Furthermore there are always strange "anomalies" which refuse to submit to being crammed into the conceptual boxes available to the normal scientist. As the life of a particular set of paradigms nears its end, these anomalies and insoluble problems tend to multiply. At the same time, the store of problems which do admit of solution by accepted methods is nearing exhaustion. As a result scientists become increasingly more frustrated and begin to relax the usual restrictions on what is or is not "scientific." The scene is now set for a "scientific revolution."

Kuhn argues that "revolution" is an apt term for the phenomenon he wants to describe. It is an overthrow of the hitherto accepted conceptual framework and its replacement by a new one. The new framework, like the old, comes embodied in some spectacularly successful piece of scientific work which solves a problem which has hitherto resisted all efforts, by the introduction of a novel postulate or point of view which makes no sense in terms of the old view. This new point of view is not derived by logic or by any formalizable process. It is a result of a creative act. The product of this act is a new paradigm which founds a new kind of normal science which will thereupon live through the same kind of life cycle as its predecessor.

A new paradigm is by definition a piece of work which is successful in the eyes of the scientific community. It must be actually persuasive to a majority, at least, of that community or they will never adopt it. In order to be persuasive, it must preserve at least a great part of the solid results of the system it replaces and add enough new results and new prospects to more than compensate for whatever losses occur. Thus Einstein's relativistic physics does not simply overthrow Newtonian physics. Within Einstein's conceptual framework Newton's system is revalued and becomes a valid and useful approximation for the case in which velocity is small compared to the speed of light. Physicists trained in Newtonian mechanics can therefore continue to use their skills to obtain valid results, with the comforting awareness that relativistic mechanics shows them to be good approximations to the "truth," which presumably is what relativistic mechanics yields.

In the enlarged, second edition of his book published in 1970, eight years after the first, Kuhn modifies his position somewhat in response to criticism.[8] He emphasizes that paradigms must be considered in relation to the community structure of science and that at least two different meanings of the term must be distinguished. The relevant community is a small group, usually less than 100, doing active research on the same problem. These men share the same "disciplinary matrix," a complex which includes formal symbolic generalizations, commitments to various beliefs and values, and finally, "exemplars," "the concrete problem-solutions that students encounter from the start of their scientific education" as well as some of the "technical problem-solutions found in the periodical literature." The exemplar, or shared example, is what the term "paradigm" chiefly intends, though at times it is used of the whole disciplinary matrix.

Kuhn's discussion corrects the impression given by the first edition that the term "scientific revolution" refers only to the great intellectual upheavals associated with names like Newton, Darwin, or Einstein. Such "revolutions" can and do occur within a small group of working scientists and remain relatively unknown outside that group. Therefore they are not so rare or so profound as one might have thought after reading the first edition. It seems that Kuhn is veering toward admitting that "normal" and "extraordinary" science are not neatly separated activities but rather two different ways of thinking which can be employed in various combinations as need arises. It remains true, however, that radical revisions of basic ideas are infrequent enough so that such events can be identified as unusual and especially significant episodes.

There are interesting parallels between Kuhn's account of scientific development and Piaget's account of individual development. The action of both the individual child and of the scientific community starts with some existing structure and assimilates new data to it. In the process of assimilation, accommodation becomes necessary. The degree of accommodation varies widely with circumstances and the particular stage of development. At the beginning of each new period in the development of the child, and during scientific revolutions when new paradigms are being established, accommodation is very prominent. A new type of organization is being created. Old structures must be modified and integrated into new patterns. The new patterns themselves cannot be stable until all their elements have been created and fitted into the proper relationship to one another. As a result the behavior of both child and scientific community tends to be erratic during these times. Once the new level of organization has been achieved, a whole new range of behavior is accessible to the child or to the scientific community, and a period of rapid consolidation and exploitation of new powers follows. During this time assimilation predominates.

The changes in mentality produced by scientific revolutions, however, are not so profound as the changes in the mentality of the child when he enters another of Piaget's periods. Even major scientific revolutions such as the Newtonian or Darwinian correspond more to the transitions between stages or substages within periods. But the total impact of the whole development of science may perhaps be equivalent to a transition to a new Piagetian period.

There are serious limitations to Kuhn's viewpoint. He characterizes his view of the development of science as evolutionary. Normal science might be compared to the successful spread of a well-adapted species into new territories. The breakdown of normal science resembles the crisis a species encounters when it enters a new and different environment or when the old environment changes. Novel ideas are like mutations. Those which are accepted and become paradigms are like successful mutations which spread throughout the species, altering its genotype, phenotype, and behavior, and enabling it to adapt to meet the challenge.

Unfortunately the kind of evolution which Kuhn has in mind is nonteleological Neo-Darwinism, and his understanding of science is limited by reductionistic blind spots analogous to those present in that account of evolution. He sees scientific "progress" as a process which results in a more complex and better articulated body of theory which can deal successfully with an increasing number of phenomena and is an increasingly apt instrument for the control of nature. He does not claim, however, that this process results in a better and better approximation to the truth and, in fact, has serious doubts about this. He apparently regards as naïve the conviction of most scientists that scientific revolutions result in something which is intrinsically "more valid," "more valuable," even "truer," than what is replaced. The creative acts which produce new paradigms are not clearly presented as anything more than random processes no more admirable than the random mutations of Neo-Darwinism. Kuhn might shrink from, but he has no coherent basis for repudiating, the statement that just as there would be no absolute loss if the species destroyed itself and left the world to the insects, so there would be no absolute loss if modern science perished and once again men understood the world only in terms of common sense and myth. It is a fact that today men understand nature in terms of highly developed scientific theories, just as it is a fact that they once understood it differently. One state of affairs is no more re-markable or inspiring than the other.

3. THE ACT OF CREATION ACCORDING TO KOESTLER

Arthur Koestler has written a significant study on the nature of creative thought.[9] Normal science is characterized by what Koestler calls "associative thought." This type of thinking operates within a certain structure (a "matrix" in Koestler's terminology) in accord with a set of rules of operation (or "code") proper to that context. Associative thought is sometimes quite rigid, but it may also exhibit considerable flexibility and strategic skill in applying the code to a complex problem. Since it is always ultimately bound to a particular matrix of thought, it is basically conservative.

Koestler's "matrices" of thought are analogous to Piaget's structures of operations, and, as we shall see, both are related to the "tacit powers" of the human person which, according to Polanyi, are the source from which all acts of articulate or explicit thought emanate. All three men see the acts of in-telligence as skillful performances which are similar in struc-ture to other bodily acts and in fact develop out of them. Koestler's understanding of a skill or matrix is so broad that he is able to speak of "the morphogenetic skills which enable the egg to grow into a hen, of the vegetative skills of maintain-ing homeostasis, of perceptual, locomotive, and verbal skills."[10]

Extraordinary science involves what Koestler calls "bisocia-tive thought," a type of thought which, if successful, produces creative insights which found new paradigms. According to Koestler it essentially involves the unification of formerly dis-tinct matrices of thought into a new and higher unity. This kind of thought is "super-flexible"—it is not bound by codes but creates a new code. It involves at least some destruction of old mental patterns, for the pre-existing matrices must be modified in order to enter into unity. Koestler compares the process to that of the regeneration of injured or destroyed tissues.[11] When a salamander's leg is amputated, the tissue near the wound first de-differentiates and regresses to an embryo-like condition. Then a process of development sets in, and a new limb is formed in very much the same way as in ordinary embryonic development. What has happened is that the full genetic potential of the cells near the wound, which is ordinarily inhibited by the presence of the developed limb, has been liberated to create a new structure.

Once the creative powers of the human intelligence have produced a set of intellectual structures adequate for successful dealing with reality (in Piaget's terminology, once a successful equilibrium has been attained), they are normally somewhat inhibited. A scientific crisis, however, breaks down the complacency of "common sense" (I refer here to the rather ethereal kind of common sense associated with complex but familiar scientific conceptual systems) and liberates basic creative powers. These powers must generally dissolve, or at least soften the rigidity, of the upper layers of the existing intellectual organization in order to carry through the creation of a new structure. The knowing subject carries out this process utilizing the lower-level structures which are not involved in the transformation, and thus subliminal or unconscious processes play a great part in intellectual creation.[12]

I believe that both Koestler and Kuhn overemphasize somewhat the destructive aspect of the creative process. The development of new structures in children does not require much destruction of those already created, though no doubt some modification is always needed in order to produce unity. But when the adolescent or the graduate student introjects great masses of developed theory, the situation is different. He is then inevitably "programmed" and is to some extent the victim of the immanent dynamism of conceptual systems which he has not fully assimilated or criticized. I am not referring here to the mentality of a mindless memorizer but to an inevitable condition which afflicts even the brightest and most perceptive modern man. It is the price we pay for the power given us by our highly developed culture. As a result, scientific revolutions and creative acts in general involve a destructive phase. However, this is a result not so much of the intrinsic nature of the creative process as of more superficial aspects of the situation.

4. PERSONAL KNOWLEDGE

At the root of the curious incompleteness of Kuhn's understanding of science is an inability to cope with the concept of truth and the nature of the human subject who knows it. Even Koestler's brilliant work on the act of creation is flawed by the same inadequacies. Koestler is explicitly antireductionistic, and he sets in bold relief the irreducible dignity of the human act which creates novelty in both art and science. But in the end he leaves this act hanging unsupported in mid-air.

The ideas of Michael Polanyi enable us to move farther. For Polanyi all knowledge, including scientific knowledge, is personal and implies self-commitment on the basis of tacit understanding which cannot be completely formalized. To know is to perform a skillful act which, like all skillful acts, is the actualization of a structure known only implicitly during the performance. A swimmer is not explicitly aware that he is maintaining increased buoyancy by holding a greater than usual amount of air in his lungs or a bicyclist that he counteracts an incipient fall to the right by turning to the right in such a way that "for a given angle of unbalance the curvature of each winding is inversely proportional to the square of the speed at which the cyclist is proceeding." [13] In an experiment done in 1958 Eriksen and Kuethe "exposed a person to a shock whenever he happened to utter associations to certain 'shock words.' Presently, the person learned to forestall the shock by avoiding the utterance of such associations, but, on questioning, it appeared that he did not know he was doing this. Here the subject got to know a practical operation, but could not tell how he worked it. This kind of subception has the structure of a skill, for a skill combines elementary muscular acts which are not identifiable, according to relations that we cannot define." [14]

A person who wishes to act skillfully must focus his attention on the act as a whole. To the extent that he focuses on details, he destroys his "concentration" and the skillfulness of the act. It is true that golfers will go to the practice tee and there include some detail of their swing within the focus of attention in order to improve it. But for maximum performance it is necessary to reintegrate the part back into the whole so that it is once again known only implicitly in the total performance.

Perceptual skills exhibit the same structure as motor skills. "Physiologists long ago established that the way we see an object is determined by our awareness of certain efforts inside our body, efforts which we cannot feel in themselves. We are aware of these things going on inside our body in terms of the position, size, shape, and motion of an object, to which we are attending." [15] Ultimately perception is a skillful act by which we construct in our nervous system a higher-order pattern, the percept, whose elements are lower-order acts of awareness of our interactions with the environment and of our own bodily efforts. We are aware of these lower-order acts only implicitly, in terms of the pattern to which they contribute. The process is seen more clearly in the case of a person who learns to use a stick to explore a cavern. At first he is aware of the sensation in the palm of his hand as his probe hits various objects. But as he acquires skill he begins to attend from his own sensations to a pattern which is potentially present in them and which he now begins to construct explicitly. Finally he has a sense of actually contacting what is at the end of the stick. The overall pattern is now the phenomenon of which he is explicitly aware and which has meaning. The sensations in his hand are known only implicitly in the higher-order whole into which they are integrated, and their meaning is grasped only in the whole into which they are projected. The stick has been so integrated with the self that it seems to be an extension of the body rather than something separate and opposed to it.[16]

To say that a familiar instrument is in some sense an extension of oneself is not an empty metaphor. The way it extends the self is really analogous to the way parts of the body extend the self. We argued earlier that only those structures whose acts are immediately integrated into conscious experience form the inner core of man. Other subsystems of the body, whose interaction with the personal center is not immediate and whose acts must be transmitted to the personal center through time and space or through intermediate levels of organization, are not part of the self in the same sense as are the former. Besides being parts of man, they are also instruments. There is a spectrum of participation in human existence which ranges all the way from immediate participation to the instrumentality of familiar tools. The degree of participation depends upon the kind of interaction between the personal center and the particular system.[17]

Intellectual skills exhibit the same pattern as motor and perceptual skills. Piaget has shown that the operations of intelligence are also skillful bodily actions. They are learned over a long period of time and presuppose developed lower-level structures. These lower-level structures are the elements which are put together into a higher-order pattern by the intellectual acts. They are the "ground" from which the intellectual "figure" emerges.

The more basic intelligibilities are constructed so early in life that later they seem immediate and intuitive to us. But this does not mean that such intelligibilities are not constructions. For example, the concept of number is a whole constructed from suboperations. Its meaning is projected into it from these suboperations. "When the child is capable of reversible seriation operations and of genuine classification (inclusion relation and all), then and only then will he be in a position to really understand what numbers are and how they behave."[18] It is not just that after learning to perform operations A and B, one becomes capable of performing operation C. Rather, just as one "feels" the floor of the cave in terms of the implicitly known pressures in the hand which holds the stick, so one understands the concept of number in terms of the implicitly known operations of seriation and classification. If the meaning of these operations is made the focus of attention, then their meaning is in turn derived from yet more elementary operations.

Thus every intellectual act is dependent upon implicitly known subacts out of which it is constructed, just as the vital acts of an organism are dependent upon the acts of cells, molecules, atoms, elementary particles, etc. Furthermore, within the unity of the intellectual act itself, the focally known intelligibility emanates from the implicitly known subject, which includes not only pre-existing and more elementary structures but also the creative self who is here and now constructing the intelligibility known.

Polanyi's insistence on the importance of the person provides a necessary supplement to Piaget's account of human intelligence. As we have seen, Piaget characterizes the stages of human development in terms of the systems of operations which the person is able to use. This is a fruitful approach to the scientific study of human intelligence, since it focuses attention on something which can be studied and measured objectively, namely, the systems of operations which the person has at his disposal. This approach should not be absolutized, however, since it omits what is most important, the essentially subjective personal center who constructs the formal systems for his own use. A man is not just an organized structure like a computer; the only thing he resembles is a man–computer system. Really, of course, it is the man–computer system which resembles him. The computer is an extension of the human nervous system created by nonbiological means.

Polanyi's emphasis on the importance and ubiquity of implicit knowledge is verified by Kuhn's study of the importance of paradigms in the life of the scientific community. Scientists do not generally agree, nor do they need to, about the philosophy of their common enterprise. They operate in terms of implicits which are assimilated with the paradigms they accept.

It is impossible for scientists, or anyone else, to explicitate completely the basis of their knowledge. Intellectual acts are the expression of the whole life of the organism, which is, in turn, the expression of the dynamism of the whole universe. When others question us, or when we question ourselves, about the basis of our judgments, we justify ourselves by explicitating enough of our implicits to arrive at something sufficiently solid and well grounded to satisfy the questioner and to make interpersonal communication and cooperation possible. However, it is never possible to explicitate everything. There is always an element of self-commitment in our knowledge acts. We rely on the essential soundness of our own implicits and also on the soundness of what we receive from those others whom we take as our authorities. This self-commitment is a kind of natural act of faith without which we could not understand anything. Polanyi quotes with approval the famous dictum of Augustine that one must first believe in order to understand.

Furthermore, just as the supernatural act of faith of a Christian believer is a free act, so the natural act of faith one makes in order to understand scientific truth is a free choice. Among the implicits which undergird our acts of understanding are desires, our "intellectual passions" as Polanyi calls them. The creative act which calls into being a new scientific theory, or even the act by which a person assimilates an established theory, is not completely determined by the data. Like the constructions of art and music, these constructions are partly determined by what the person wants. Scientific creation is therefore a moral act, an act which expresses the love of beauty and order which motivates the scientist. In his famous essay on the psychology of mathematical invention, Poincaré suggests that the mathematician's sense of beauty is one of the most powerful enabling factors in the creation of new mathematics.[19] This sense of beauty clearly involves the will's attraction by the intellectual good. The same is true in all fields. Men create the kind of world they want. Intellectual acts are not only acts of understanding, but also acts of love, though the relative weight of the two components can vary greatly in different fields of endeavor.

Furthermore, without an understanding of the human person, there is no explaining those creative acts which establish scientific paradigms and create a community of scientists able to cooperate in studying nature. Kuhn ends by comparing them to the random and (according to the Neo-Darwinians) ultimately senseless mutations which carry biological evolution forward. But nothing comes from nothing, and from senseless accident only a senseless kind of science could come. Unless there is such a thing as truth, a value known to human persons who practice science, and unless scientists out of love for it seek and find it, the whole enterprise (like all other human enterprises) is pointless. Koestler sets the importance and dignity of the creative act in bold relief and says much which is illuminating about its nature. But in the end he leaves it hang-

ing in mid-air like the smile on an invisible cat. Whence does
it come and what does it aim at? Koestler does not know.
Polanyi has begun the investigation of the essential questions.
Scientific knowledge, like all knowledge, is personal. It is the
act of a human person who understands implicitly far more than
he ever expresses within the formal scientific disciplines and
who acts out of personal intellectual passions and social instincts which neither arise from nor can be justified by science.

Again, unless the great men at the origins of scientific revolutions, as well as the many ordinary men who accept them,
possess a subjective heuristic anticipation of truth, there would
be no accounting for the fact that these upheavals produce
progress. Sheer change is more likely to do harm than good unless it is guided by some intimation of a better reality. Kuhn,
who apparently is not sure of the existence of truth, ends by
doubting if scientific progress is really progress in the way in
which most scientists suppose. This is quite logical, though mistaken. Reductionistic scientists, on the other hand, are quite
right in thinking that they are making real progress, but illogi-
cal in not rejecting their reductionism. The two are incompatible.

What we call "truth" is, then, freely chosen by the person
for unspecifiable reasons. Does this mean that it is an arbitrary
construction and that the old definition of logical truth, "conformity between the knowing mind and the reality it knows," is
not correct? If one freely creates the mental structures one
wishes, it would seem that their conformity with external reality
is a matter of accident and not a defining property. Polanyi
does not seem to have resolved this question completely. The
same is true of Koestler and of Kuhn, who closes his book with
the assertion that this and related questions have never been
answered.

Since the breakup of the medieval world, Western man has
been haunted by the question of truth. Under the double impact of the Reformation and the rise of science, old certainties
dissolved, and he found himself faced with the problem of
working out a viewpoint on a changed world which would be at
least partially new and yet *true*. Descartes, Leibniz, and
Spinoza, men filled with faith in reason, hoped to do this by

clear and rigorous thought. Hume concluded that they had not, that he could not, and that very likely no one could. Kant grimly resolved that he would do his duty and cling to faith in God and morality while trying to salvage the beautiful and fascinating world of Newtonian science from the whirlpool of doubt. What has Polanyi done? He has told us, in concert with most modern philosophers, that the questions about truth and about the human person belong together. Along with other contemporary philosophers, whose ideas we shall discuss shortly, he has shown that tacit knowledge and heuristic anticipations are crucial elements in the discovery of truth. Perhaps his most constructive contribution has been to show that scientific knowledge stands in essentially the same relationship to truth as the rest of human knowledge. By destroying the illusion of "objectivism" he forces scientists, and the various schools of philosophy and utopianism loosely associated with science, to face the fact that they are involved in the same situation as artists, moralists, theologians, philosophers, and ordinary people. This is important because it simultaneously cuts down the "scientistic" pretensions of those who want to deify science and the "humanistic" pretensions of those who want to degrade it. Finally, he has affirmed the fact that the human mind can and does attain truth. Polanyi admits that he does not quite know how. Nevertheless he appraises his own implicits in the very act of making certain judgments and finds them valid. He commits himself to their truth, believes them, and tries to persuade others. This is the act of a mature and responsible member of the human community.

It is important to realize that self-commitment to the truth is also an act of self-affirmation. The act of making a true judgment is a skillful act. Only those who have the necessary skill can do it. But self-confidence as well as skill is required. The golfer who steps to the tee worrying about how he will control and coordinate the thousands of muscles involved in his swing is unlikely to produce a good shot. Doubts about whether there is such a thing as truth or about whether man can attain it are simply pathological in spite of the fact that intelligent and learned persons have entertained them. Only those who are healthy minded enough to maintain their equilibrium while they philosophize and thus to continue to know that they are in contact with reality will ever be able to produce a good theory about how they do it. To return to the example of the golfer: the successful teacher is the one who can maintain his well-integrated swing even while he probes into its details so as to become explicitly aware of them and thus capable of transmitting to others in explicit terms what he already knows implicitly. If his implicit knowledge of golf is destroyed by the attempt to make it explicit, he will be neither a golfer nor a teacher of golf.

The problem before us then is not, can we attain the truth, but rather how do we attain it, and how can we attain it more fully? Polanyi has shown that the answer is related to the nature of the human person and of implicit knowledge. We must inquire more deeply into these.

NOTES

1. For an overview of Piaget's work see FLAVELL, FURTH, PIAGET (I), (II).

2. There seems to be a conflict here. The functional invariants characterize the nature of intelligence, yet they also characterize biological functioning in general. The apparent conclusion is that intelligent life is not essentially different from what preceded it. In other words, Piaget is a reductionist. Yet there are abundant indications elsewhere that he is not a reductionist.

The explanation seems to be that Piaget, in spite of frequent statements to the contrary by his commentators and his own early interest in philosophy, is not really philosophically inclined to any notable degree. Of course it is true, as even his admirers insist, that his terminology is shifting and unclear. But beyond that, he seems to overlook problems and to tolerate confusions which no good philosopher could. To use a comparison: he reminds me much more of Lord Rutherford than of Einstein. Einstein was a theorist who never displayed much interest in doing experiments, and Rutherford, a great experimenter who invented theories to make his experiments intelligible. Rutherford stayed close to the facts, and this practice was both a source of power and a limitation. Piaget will use general (some would say philosophical) ideas to guide his experiments and to explain his observations as, it seems to me, any really good psychologist must. In contrast with his overly positivistic colleagues he may appear philosophical, but I feel that, in spite of his remarkable abilities which make him one of the outstanding thinkers of this century, he is not a philosopher.

3. FLAVELL, p. 86.

4. FLAVELL, pp. 211, 212.

5. FLAVELL, pp. 78, 79.

6. See FLAVELL, p. 80.
7. DÉCARIE.
8. See LAKATOS also.
9. KOESTLER.
10. KOESTLER, p. 415.
11. KOESTLER, p. 454.
12. This explanation is not Koestler's. In fact it seems that he would be incapable of giving it since he does not recognize the full meaning of the concept of "knowing subject."
13. POLANYI (I), p. 49.
14. POLANYI (III), p. 8.
15. POLANYI (III), p. 13.
16. POLANYI (III), pp. 12, 13.
17. See Whitehead's definition of the human body as "that region of the world which is the primary field of human expression." WHITE-HEAD (III), p. 30.
18. FLAVELL, p. 312.
19. POINCARÉ.

5

Symbolizing Activity

1. THE PRIMORDIAL HEURISTIC ANTICIPATION

WE HAVE SEEN that at each instant of time a new increment of reality emanates from the past so that the comprehensive present includes the whole of past time plus the objective present plus the action by which the former is giving rise to the latter. This general pattern applies to the whole cosmos and within it to every entity which is still developing. Structures which originated in the past are on the subjective side of the action; those which now exist for the first time are on the objective side. If the action in question is on the human level, it has that quality of self-transparency which we call consciousness. The human subject, which includes all the structures already established in him by his past, consciously produces new knowledge and new decisions. The new knowledge and decisions are objectively, focally, known. The old ones are subjectively, tacitly, known. This is true even when we recall our past acts. The past acts are the source from which emanates a new bit of knowledge, our present memory of the past act. Thus the past act is known subjectively in the performance of the present act and also objectively, but indirectly, insofar as it is symbolized by the memory now being produced.

103

Subjective knowledge never ceases to be subjectively known, though it may also become objectively known in the symbols emanating from it. There is a good deal of variability in the noetic structure the person has from moment to moment. At each instant the subjective aim of the person lays hold of the various elements of his past and integrates them into a configuration suited to his purpose. This act of integrating the elements of the past into a currently relevant noetic structure is nothing other than the act of symbolizing this structure. It is in producing the symbol that the person organizes the subjective structure from which the symbol proceeds. Thus one's subjective awareness of one's past varies from moment to moment. A given structure will be focal or peripheral depending on the way in which it is contributing to the production of the instantaneous present.

Some subjective knowledge can easily be made objective. Thus the implicit knowledge which the touch typist has of the location of the keys can easily be objectified if he wishes. Other kinds of subjective knowledge are difficult to make objective. Anyone who has seriously tried to change his moral character or his psychological structure is aware of this. Nevertheless, on occasion one can become aware that an attitude or a way of viewing things which seems simply given, a part of the very nature of things or at least of one's own nature, is actually a decision or a conclusion, something which was adopted or constructed on a given occasion for definite reasons, and which has been a part of tacit self-awareness ever since. Furthermore, it seems that under the impact of crisis, disease, drugs, or special training, elements of the self which ordinarily are never the object of explicit awareness can become so. Nevertheless, there is an essential core of implicit knowledge which can never be explicitated adequately. We shall argue later that the essential self is not a structure in the ordinary sense of the term. It is the source of all the structures of the human personality, which it produces and integrates into itself by its acts for its enrichment and self-fulfillment. But the essence of the human person transcends the created universe and cannot be produced by anything within the cosmos; it is identical with the implicit knowledge and desire which constitute the primordial heuristic anticipations guiding the entire quest which is human life. We must now inquire into these essential and universal implicits which found all human knowledge.

"Man by nature desires to know." [1] He is the wonderer, the questioner, who is never satisfied with the understanding he has attained and for whom any answer is the occasion of further questions. Aristotle, Piaget's children, and the modern scientist are all brothers because they are never satisfied with what they know. In this they differ essentially from contented cows, cats, dogs, and even chimpanzees. A close consideration of the act of questioning will tell us a great deal about the nature of the human desire to know. [2]

This act reveals, first of all, that the questioning subject does not know, and yet at the same time knows, what he is inquiring about. Clearly he does not know the precise object of inquiry explicitly or else he would not ask. Yet at the same time he must know it implicitly or he would not be able to ask. Here we have the basic insight of Plato in the *Meno*. Plato's answer in terms of "remembrance" is inadequate, but the insight remains. The unknown which one seeks must already be known or it could not be sought. The implicit knowledge one has of the explicitly unknown object which one inquires about has been called a "heuristic anticipation." In science and mathematics such anticipations are extremely important. Without them no solution to a problem can be constructed or recognized as such when found. Indeed, it can be said that the formation of correct heuristic anticipations is the better half of truly creative scientific work.

The heuristic anticipations of the scientist, and of the mature person in general, presuppose a great deal of knowledge which has already been acquired. Adults always inquire about something which they already know explicitly, though partially. It is their already-acquired understanding which enables them to have complex, well-articulated, and appropriate heuristic anticipations. Though mature questions presuppose mature knowledge, still this knowledge is not the core of the act as such. An act of inquiry does not presuppose any particular explicit knowledge. Even if we fall into doubt about all that we know, we can still inquire. If we doubt our right to ask a question, we can inquire about this too, and the question about the question has the same essential nature as the original question. The essential implicit knowledge which underlies all questioning cannot, therefore, be tied to any specific object, situation, or explicit understanding. It must be a general knowledge which gets specified in different ways in different situations.

Every act of knowledge includes an element of contemplation as well as an element of questioning. It is contemplation insofar as we find rest and enjoyment in the understanding we have achieved, and inquiry insofar as our drive to know strains beyond what we have already attained toward the unknown anticipated in our implicit knowledge. In some acts, the contemplative aspect predominates; in others, the aspect of inquiry predominates.

Inquiry implies manipulation of the environment in order to elicit further information. We may address a verbal question to an authority, or we may combine our own concepts in a new way in the hope of achieving greater coherence. In any event the person's drive to know affects his "world," i.e., the other which confronts the subject, and this includes the structures within his own body insofar as they are objects of knowledge and manipulation, as well as what lies beyond his skin. Thus the basic and universal heuristic anticipation is not cool and passionless. At every moment it results in action. When one questions, one strains with desire beyond what is already known. The person's basic anticipation is therefore a desireful anticipation, or else an anticipatory desire, of all that can be known.

2. THE NOTION OF BEING

The nature of the basic anticipation can be seen more clearly if we ask ourselves what a person can inquire about. We can inquire about anything which has, or can be conceived as having, being. Only that which does not exist in any sense escapes inquiry, and that is nothing. So the act of inquiry as such has no limits. If anyone should claim that it does, we can inquire about those limits, thereby showing that they too are included within the field of inquiry, which extends indefinitely beyond them. This means that the basic heuristic anticipation which makes possible every act of human understanding is an anticipation of the totality of that which is or which can be, of all possible and actual beings. Clearly it is not an anticipation of them according to their particularity. A mature question about the specific nature of anything requires mature knowledge. Instead, it must be an anticipation in terms of what is common to them all, namely, being. The implicit knowledge of being provides the structural framework, field, or horizon within which every real or possible entity can be understood. Each of them is a particular realization of that general intelligibility, a limited way of having being which is both related to and distinguished from other particular existents.

The notion of being cannot be defined directly. In the first instance, it is implicit and unformalized. But we have succeeded in saying something about being already and shall say more. Therefore we must have some kind of explicit knowledge of being. We do, of course, know something about particular beings such as electrons, men, unicorns, and topological spaces. Knowledge about beings is, in a sense, knowledge of being. But this knowledge does not enable us to grasp explicitly the general intelligibility which they all share and which enables us to call them beings. We do not know this general intelligibility directly and immediately.

But we can know being explicitly in two ways. First, we can formulate an explicit heuristic anticipation through a reference to our own acts of knowledge and desire. Thus we say that being is the totality of all that can be known or desired. Secondly, we shall arrive in the end at an explicit notion of being on the basis of reasoning and revelation. One might say that this whole book is an explicit definition of being. It is not an immediate and intuitive one, however, but a remote and complex conclusion based on implicit knowledge of being, explicit knowledge of the particulars of being, and a great deal of reasoning.

Thus if we mean by "definition" "a word or phrase expressing the essential nature of a person or thing," [3] then being cannot be defined. The notion is implicit knowledge, or an explicit, but very general, heuristic anticipation, or a mediate, complex, higher-order intelligibility.

I accept the argument of Lonergan that all the chief metaphysical categories are of this nature.[4] This is obvious in the case of the other "transcendentals," viz., truth, goodness, and beauty, which are identical in re with being and differ only by connoting a particular relationship to ourselves. It is true also, in a modified sense, of such notions as causality, symbol, object, subject, material, actual, spiritual, relation, etc. All these are ultimately contained in subjective self-awareness and therefore, like being, can be known in three ways: as implicits, as explicit heuristic anticipations, or as mediate, complex, higher-order intelligibilities.

Metaphysics is precisely the discipline whose task it is to make explicit the fundamental heuristic anticipations given with the very essence of the human spirit which we grasp implicitly in every human act. Therefore metaphysics is the most general and fundamental of the knowledge disciplines. It is also, in a sense, the most useless one, for it establishes nothing which is not already known. As we have seen, a physical scientist does not need metaphysics in order to practice his science. In the abstract, even a social scientist does not absolutely need it. He has the implicits available, and if need arises, he can explicitate them to the extent required in order to solve his problems.

In practice, however, things do not work out that way. Without a healthy discipline of metaphysics to inject correct explicitations of fundamental implicits into the cultural atmosphere, physical and social scientists frequently arrive at conclusions which are a strange mixture of science and bad metaphysics. Neo-Darwinism as commonly presented is a good example. This, of course, is not particularly insulting to physical or social scientists. Metaphysicians as well frequently arrive at strange conclusions. Metaphysics is a difficult and risky enterprise. But the fact remains that one of society's greatest needs today is a sound metaphysical tradition, sustained in being by good metaphysicians and absorbed by osmosis into the intellectual bones of the general population, especially the intellectuals. I might add that I think this is impossible without a healthy theological tradition which would stand in somewhat the same relationship to metaphysics as metaphysics does to the other sciences and without a religious conversion which would make a healthy theological tradition possible. In the end, the health of the intellectual life of the human race depends on prayer, contemplation, and, above all, love of God and neighbor.

3. THE TRINITARIAN ORIGIN OF BEING

The history of Western philosophy began with the attempt by the Greeks to understand the "origin of everything." "Origin" is used here more in a causal than a temporal sense. The question is not so much about the sequence of events in the history of the universe as about the (unchanging?) structures of reality which are responsible for all events, whatever their place in the sequence. Plato, the first of the supremely great Western philosophers, answered that the phenomena we experience with our senses are mere reflections of an invisible realm of eternal, necessary intelligibilities, the "Ideas." It is this realm which is "really real," and the phenomenal world of ordinary common-sense understanding derives its reality by participating in the higher reality of the Ideas. Aristotle accepted Plato's emphasis on the importance of the suprasensible but insisted that essences are in, not above, material things and that we understand the divine through the sensible rather than by intuition. For him metaphysics was the science of "beings as beings." Thus, its object is every thing which exists, visible and invisible. Metaphysics investigates every thing, insofar as it has reality, and therefore must strive to understand both the intrinsic and the extrinsic causes of things, particularly the suprasensible and divine realities which lie behind what is visible. St. Thomas Aquinas developed and applied the concepts of Plato and Aristotle to Christian theology. He emphasized particularly the distinction between beings and their "to be" (esse), the act by which they exist. The pure act of being is fundamental in his philosophy.[5]

The beginning of the modern era in philosophy was characterized by the loss of the insight into the importance of being and by a renewed interest in the nature of knowledge stimulated by the great increase in scientific understanding.

Kant realized that the kind of theoretical understanding attained by science cannot be explained simply in terms of the atoms of information obtained from individual experiments. In modern terminology, he saw that the information contained in a message is determined by the "message space" in which it exists. Consequently he set out to find the universal *a priori* conditions which underlie all human understanding. Probably the main factor contributing to his failure was the fact that, as a result of the loss of insight into being which had occurred before his time, his inquiry was too much confined to essences and formal structures. In the twentieth century this insight has been regained by the existentialists and neo-scholastics. Furthermore, for the first time, the importance of the person has been grasped with full clarity, and, as a result, the stage has been set for an advance to a new level of philosophical understanding.

I believe that this advance has actually occurred with the recognition by Karl Rahner that the traditional Christian understanding of the Trinity provides us with a key for the understanding of many other things as well. In fact, Rahner's concept of being and of symbolizing activity has such remarkable power that it is able to serve as a foundation for a comprehensive view of the universe which goes beyond the field of theology in which he is primarily interested.

This conception is not intrinsically dependent on revelation. I doubt that it would ever have been stated were it not for the existence of revelation, but once thinkers have understood the idea, it can be grounded in considerations independent of revelation. It is therefore a philosophical idea rather than an essentially theological one. (The knowledge that the prime analogate of the conception is found in God is, of course, theological.) Nevertheless, in the remainder of this book I shall appeal directly to theological data. I believe that much of what I have to say could be established without such appeal, but to do so would be tedious and time-consuming and, above all, would not display the essential structure of the ideas and of the cosmos, which cannot be fully appreciated without being seen in relation to God's gracious love.

The God of the Old Testament Israelites, the maker of Heaven and Earth, is a transcendent being. He has created all things out of nothing, and He is their absolute master. His power is limitless, for He calls things which are not as though they were, and they, being nothing but for Him, cannot limit Him in any way. Furthermore though He loves all things He has made and takes the most intense interest in their every particularity, He has no need whatever of them. All the goodness and beauty found in creatures exist in them because of Him, and if He did not possess it already, it would never have become theirs. He creates not out of need but out of pure goodness and generosity, and were the world *per impossibile* to escape His sustaining hand and to return to nothingness, He would be none the poorer. He is the "necessary" being, the one who is conditioned by nothing and whose act of existing is rooted in Himself alone. His proper name is "He who is." [6]

This God is ultimate mystery itself, the explanation of everything, who Himself can never be explained or fully understood. He is the light so bright that it cannot be seen by mortal eyes, and the truth so clear and so full that it cannot be fathomed. He is the fullness of being, the ground of every value and perfection. Man knows far more clearly what He is not than what He is and can speak positively of Him only in terms of analogies which point to what is not seen.

We, for our part, are contingent beings. Our existence is not ultimately from ourselves, though we do exercise it, but rather from God who calls us out of nothingness. It is only because His hand guides and sustains us that we are able to perform ourselves at all. Our truth and love are at best flickering shadows of His, and our freedom is never far from slavery to fate, sin, and death.

But in the New Testament we encounter a man like no other, Jesus Christ. He is fully human, possesses human consciousness, human emotions, human intelligence and freedom. He lives, suffers, and dies, as all men must. Nevertheless the apostolic witnesses who lived with Him during His brief life came to believe that there was something more than the merely human in Him, and this belief has been expressed and transmitted in the New Testament writings.

The "search for the historical Jesus" and the *ipsissima verba Christi* is at best an arduous one, and the extent to which even the best scholars can succeed is problematic. Hence, it is difficult to determine precisely the evidence which led to the conclusions at which the apostolic generation of Christians arrived. Nevertheless, the conclusions themselves are clear. Jesus Christ is from "above." Though He was born of a woman in Palestine during the reign of Augustus Caesar, this same person is the Word of God who was with God in the beginning. Before Abraham or any human reality came to be, He is. Through Him all things come to be, and in Him, they are held in unity and find their meaning and purpose. Finally, it is in and through Him that creation will reach its ultimate goal, God the Father. The relationship of Christ to God is expressed in many different ways by the New Testament. The metaphors which have been most influential in the development of theological understanding of the Trinitarian relationships are probably "the Word of God," "the Image of God," and "the Son of God."

In the fourth century, the Church was forced by the Arian controversy to think out and state more explicitly part of the meaning of the concrete and image-rich New Testament language about Jesus Christ. On which side of the gulf between the finite and the infinite, the contingent and the necessary, does He belong? The answer of the fourth-century councils was that this person belongs on the side of God. The Lord Jesus Christ is the only Son of God, begotten, not made, of the Father before all ages and of one substance with Him.[7]

This explicit recognition of the divinity of Christ resulted in a similar recognition of the divinity of the third divine person, the Holy Spirit. Beyond the created order, there exists not just the one God of the Old Testament but two other equally divine persons, the Son and the Spirit.

Throughout the ages Christian theologians have tried to fathom the depths of the mystery stated in the scriptures and restated for later ages by the early councils of the Church. Christians could not simply accept a multiplicity of divine persons after the manner of the pagans. Their Old Testament background required them to insist on the unity and uniqueness of the divine. Furthermore, each of the three divine persons is infinite and unlimited. But how can there be three infinites, three omnipotents? This problem peculiar to Christians has forced them to face with special earnestness and intensity of effort the ancient dilemma of "the one and the many" and to develop a deep and penetrating appreciation of relational reality.

The Father is the unoriginated origin from which comes everything else which exists. He both constitutes and expresses His own infinite reality by generating another who is like to Himself, the Son. As a person, the Son is distinct from the Father; in fact, the very personalities of both Father and Son are constituted by their mutual relationship (and by their relationship to the Spirit). Yet the Son is not separated from the Father in anything but shares with Him one life, one understanding, one will and desire. Because He is the adequate expression of the Father, He is His symbol, His image, His self-revelatory Word. He who sees the Son sees the Father from whom He can never be separated and who is present in Him as the source of all that He is and does.

In like manner, there proceeds from the Father and the Son as from one principle a third infinite person, the Spirit, who is the symbol and expression of their life in mutual union. The personality of the Spirit is constituted by His relationship to the Father and Son in their unity, and their personalities are in turn constituted by their common action of breathing forth their Spirit. The Spirit is not separate from the Father and Son in anything but shares with them one life, one understanding, one will and desire. Because He is the self-expression of Father and Son in their union with one another, He is appropriately thought of as the Spirit of love, the totally adequate expression of love, and, hence, the Gift par excellence of the Father and Son to man.

Because the total reality of the three infinite persons is constituted by their relationships to one another, each contains in Himself the whole reality of the others. There is, then, just one single infinite divine reality, one being, one intelligence, one will, which is possessed differently by each of the three.

This traditional Christian understanding of the divine is clearly quite different from the picture one would arrive at by a certain kind of philosophical reasoning. Finite beings are not totally constituted by their relationships to one another, and these relationships are therefore ontologically subsequent to their substantial existence. These relationships are actually means of filling out and perfecting what is in itself limited and imperfect. And even when a finite being has entered into all suitable relationships with other finite beings, it still remains limited and imperfect. It is thus easy for a philosopher to conclude that multiplicity and real relationships are essentially connected with limitation and imperfection and to think that the divine must be some sort of monad without internal multiplicity or distinctions and without real relationships with anything.

In actual fact, however, an infinite person can only exist as the term of a relationship by which He is totally constituted, and from this relational character of His existence, there follows both His total distinction from and His complete unity with the other divine persons. Multiplicity in unity is not antithetical to, but proportional with, the degree of being.

It should be noted that the development which occurs within the Trinity is not temporal. The existence of the Father does not precede the existence of the Son temporally. Since the entire being of the Father is constituted by His acts of generating the Son and spirating the Spirit, there cannot be a moment when He exists and they do not. Even though He is prior to them by nature, all three persons are equally eternal and without succession of moments (as far as their essential nature is concerned). But a finite entity like man, or like the whole cosmos, does not possess itself totally. Our consciousness of self does not grasp fully the whole potential of the being which we have received. Consequently we are not able to express ourselves in one exhaustive and timeless act. Rather, each partial self-expression enriches us and enables us to grasp and express in another subsequent act a little more of what we potentially, but not actually and personally, are. Temporality is imposed on finite beings by the imperfection of their self-possession.

4. THE THEORY OF SYMBOLIZING ACTIVITY

The theory of the Trinity contains a theory of being, of causality, and of symbolism. This triplex theory can be generalized into a universal explanatory principle.

There are three fundamental modes of being, three different ways of existing—that of the Father, that of the Son, and that of the Spirit. The Father is unoriginated origin; the Spirit is originated without originating in turn; the Son is originated and is the origin of another. Our existence is like that of the Son. Like Him we receive our existence and power of action from another. The same is true of other finite beings. They are all active with a power which is dependent on their previous actualization.

We can make a general statement which applies to the being of each of the three divine persons, namely, that in each case "to be" is to be the term of relationships by which one is constituted. The very concept of relation includes two terms and an orientation of one (the subject) toward the other (the object). In the case of a real constitutive relationship such as we are dealing with here, the *relata* are not external to one another but present within one another as constitutive factors. Yet this union of the terms does not diminish their distinction from one another; it grounds it.

Since the Trinity is the archetype of all being, the same statement must also apply to finite beings. "To be" is to be the passive term of a fundamental creative relationship with God and of active and passive relationships with other finite entities by which one is situated in the web of the cosmos. The whole material universe is united into a coherent system by mutual constitutive relationships. Each part is continually acting upon other parts, thereby bringing forth within them expressions of itself by which it is truly present within them and united to them even as by the same action it is distinguished from them. The unity and multiplicity of the universe are mutually dependent and supportive of one another in analogy with the Trinity, the archetype of all reality.[8]

Being and causality cannot be separated. Causality has to do with being and can be used in as many different senses as being itself. In every instance, to cause is to confer being of some type. The causality of a finite entity must be understood by analogy with that of the second person of the Trinity. It is not exercised independently. Rather, the finite entity is endowed with power in virtue of its constitution by the Father and becomes the co-cause of another by relating itself to Him.

Thus, to exercise causality is analogous to the spiration of the Spirit by the Son in the Trinity. This explains how it is that our finite power is able to cause being: because the Father works with us, leaving us free and independent just as He leaves the Son free and independent, but joining with us in the production of the effect. The nature of the effect is thus an expression both of the Father and of the finite cause. It is indeed mysterious that God can cause something without determining or knowing fully in advance what is going to eventuate. It is not, however, a contradiction, and we can gain some feeble light on this by noticing the analogy with the spiration of the Spirit, which is ultimately a mystery of love. The finite being reaches out beyond itself in desire of the final fulfillment of itself which is, whether it knows it or not, union with the Father. From this union in desire with the First Cause, there emanates another finite effect which expresses the nature of both partners. In its being, goodness, truth, and beauty, we see reflected the nature of God; in its finite essence, its particular situation in the web of the cosmos, we see reflected the nature of its finite causes.

One can ask why, if finite causation is indeed analogous to the spiration of the Spirit by the Son, the effect produced is another finite reality which is again analogous to the Son rather than to the Spirit. The answer would seem to be that this is a result of the finiteness of the effect. The Spirit is the term of an infinite act of mutual love between Father and Son and, so, is a total expression, the complete fulfillment of all possible power to love. There is no need for Him to do anything further because He Himself is the fullness at which action aims. But a finite cause is incapable of total expression. The effect it causes expresses something but not everything, and thus the effect is required to cause in turn so that step-by-step the full expression of the potentiality of the cosmos may be attained.

Besides this theory of being and causality, the theory of the Trinity contains a theory of symbolism. The Son is the image of the Father. Similarly, the Holy Spirit is the symbolic self-expression of the Father and the Son united in mutual love. Since the source and archetype of all reality is essentially symbolic, the same must be true of every thing which exists:

"all beings are by their nature symbolic, because they necessarily 'express' themselves in order to attain their own nature." [9] Thus

> each being forms, in its own way, more or less perfectly according to its own degree of being, something distinct from itself and yet one with itself, "for" its own fulfilment. (Here unity and distinction are correlatives which increase in like proportions, not in inverse proportions which would reduce each to be contradictory and exclusive of the other.) And this differentiated being, which is still originally one, is in agreement because derivative, and because derivatively in agreement is expressive. [10]

We are speaking here of that type of symbolism which is ontologically primary, not about more or less arbitrary symbols which are in large part connected by mere convention with what they symbolize. In this latter case we have two realities already intelligible in themselves which happen to agree with one another in some way so that one happens to be capable of standing for the other if men choose to make the connection. In the case of a real symbol, however, each of the two realities is at least partly constituted by the other. That which is symbolized brings forth the symbol for its own fulfillment, and the symbol exists as such precisely by completing and fulfilling the symbolized. To the extent that the two stand in this relationship, neither can exist or be understood without the other.

In the order of reality the prime analogate is the Trinity. In terms of immediacy of understanding, the prime analogate of symbolizing activity is the human knower bringing forth his knowledge. The New Testament itself hinted about an analogy between the two when it described Christ the Lord as the "Word of God" and the "Image of God," and the theory of the Trinity was developed by a dialectical process which worked back and forth between human experience of intellectual activity and the revelation about the nature of the Trinity. Now that the connection has been made explicit, we can use our immediate personal experience of knowing to give empirical content to the abstract theory and at the same time we can use the theory to give order and clarity to the experience. It is for this reason that, although the notions of being, causality,

118   COSMOS

and symbolic activity mutually contain one another and consequently have the same content, that of symbolic activity gives
easiest access to the inner meaning of all three.

5. KNOWLEDGE AND DECISION AS SYMBOLIZING ACTIVITY

Man is a questioner, an inquirer who actively approaches reality
with a noetic structure, a complex of desires and heuristic
anticipations. At the first instant of life, this noetic structure
is simply the primordial being of the human person, the implicit desire and knowledge of God and the world which is the
objective term of the constitutive relationships he has with
God and cosmos. Only within the field of inquiry, or horizon,
of this noetic structure can beings appear to man. As we have
seen, this field is simply unlimited. It can be filled, or fully
actualized, only by God Himself together with the totality of all
real and possible beings seen in relation to His creative will.

Man is already a being at the first instant of his existence; in
the second instant, he acts—that is, a symbol of his being proceeds from him. In accord with what we have already asserted,
this symbol is the objectively present reality of his second instant of life. Since it is symbolic and representative it is
knowledge in a certain sense. It conforms to the primordial
being from which it springs, and since that being is the term of
constitutive relationships, it is identical with the subjects of
those relationships. It follows therefore that the newly produced
symbol represents and is conformed to the external reality by
which he was constituted in his first instant. However, this
symbol is not entirely immanent to man. It is a complex reality
which in part remains within him and in part spills over into
the outside world. In the term produced by immanent action
the subject possesses himself and the external reality by which
he is constituted. Transient action, however, is partially lost
to the subject. Its term is external to him and enriches the environment. This, of course, is of benefit to the subject too, but
indirectly.

At each succeeding instant, a new symbol proceeds from
the entire past of the subject. Since this past is ever growing
in ontological richness, the symbols proceeding from it also

grow in richness. Thus man develops and enters into ever fuller possession of himself. His action becomes more complex, more responsive, more adapted—in Piaget's term, more "equilibrated." This is to say that the original potential contained in his constitutive relationships to God and the world are progressively actualized. Eventually this development carries through all the Piagetian stages, and man, intellectual in a broad sense from the very beginning, becomes capable of operations which are more properly intellectual.

Symbolizing activity is volitional as well as intellectual. Understanding and decision are two inseparable aspects of the same act. In bringing forth the symbol of what we (implicitly) know, we take a stand toward it, put ourselves into a certain relationship to it. In other words, we decide whether this thing is good or bad for us, whether we love it or hate it. This is so even in the most "disinterested" contemplation. Whenever we understand anything, we understand it first and foremost as a being, as something which is included with us within a common horizon of being and intelligibility and which is therefore necessarily related to us. To contemplate it is not only to know it as something distinct from us but also to know it as related to us. This relationship is either one of harmony and peace or of incompatibility and disquiet. The same external object will be symbolized and known quite differently depending on whether we love it or hate it. This is obvious, of course, in the case of other human beings. We cannot really know another person unless we love him. This is true even in the case of someone like Hitler. The one who knows such a person best is the one whose hatred of his deformities is situated within the context of a more fundamental love of his basic humanity and a painful awareness of the way it has been perverted by evil. The same thing is true in a less obvious way even of scientific or mathematical knowledge. The knowledge of physical or mathematical structures cannot be separated from an appreciation of their beauty. In fact, as Poincaré suggested,[11] those who do not appreciate them will not be able to discover them or to understand them well. The person who "hates math" brings forth a distorted and imperfect symbol of the reality with which he is in contact.

At the human level, symbolizing activity is free and creative. Its ultimate determination comes from the person himself. God and the external world determine and confer the fundamental passive being which the act of symbolization presupposes. But how this fundamental reality will be symbolized is determined not from the outside but from the inside. A human person is creative in a real sense. He creates his own personality and is ultimately responsible for it before God and man.

Human creativity is, of course, limited. The implicit knowledge and desire conferred upon us in the beginning is the horizon within which all our activity is situated. Furthermore, the possibilities of each moment are determined by the environment. We are constituted by our relationships with it. Finally, our history is a part of us. The next moment of our existence emanates from our personal center as containing and determined by the acts of the past.

But in spite of all limitations man is free. He is a genuine creator, and this makes him an image of the divine and sets him apart from all nonintelligent animals.

6. TRUTH

If the symbolic structures man creates are an expression of his attitudes, what is to be said about their truth? The classic definition of truth is that it is a relationship of conformity between knowledge and the reality it symbolizes. I shall call this kind of truth "logical truth." It is the truth proper to the λόγος or word which man brings forth within himself first of all and then communicates to others. Logical truth can come into existence only because there already exists a prior "ontological truth." This truth is the self-transparency of spiritual being which is aware of itself as a symbol, the term of the self-expressing and self-fulfilling act of the ground from which it arises. It is at once symbol (λόγος) and being (ὄν), and there is no question of error with respect to it. It simply is or is not.

Ontological truth "ought" to develop into explicit, logical, truth, but this will happen only if the man in whom this intellectual emanation takes place acts rightly. There is more than one valid mode of expression of ontological truth possible at a given instant, so that human freedom determines what aspect of the truth becomes known. It is also possible for fundamental ontological truth to be perverted by an evil decision, so that the symbol which is produced is not a true representation of the reality of the person who brings it forth. In other words, man can freely introduce into himself a split between his fundamental ontological reality and the symbolic structures he creates. He thus becomes a liar, in the fundamental sense of one who is not true to himself and his own immanent law of development. As a result, man's knowledge is distorted and partially false, whether or not he explicitly knows it. All such self-betrayals are acts of hatred, refusals to accept what one has been given in one's own existence. The order of things contained in implicit knowledge and desire is not accepted and expressed. Instead a part of it is suppressed or distorted.

Thus love of what is is inseparable from knowledge of it. This is not to say that one cannot know and hate something. But the hatred associated with true knowledge is contained within a greater love for the sake of which we accept even that which we would prefer were otherwise. The world as it is is not perfect, and so our attitudes toward it are divided. But only he who fundamentally loves and accepts it can know it truly.[12]

The question of truth is related to the question of "proof." The modern period has been characterized by a desire for clear and distinct ideas, for impersonal and machinelike decision-processes by which one can distinguish valid notions from invalid ones without risk of error or the necessity of personal integrity. One of the earlier consequences of this attitude was the realization that no one can prove the existence of God; one of the more recent ones has been the realization (generally somewhat horrified) that no one can prove the validity of scientific theories or of anything else.[13] Today a person imbued with the modern attitude toward proof cannot be certain of anything on the explicit level.

It is clear that the root of the problem is an impractical notion of proof. Certainly one can define "proof" any way one wishes; but the price of defining it in the way many contemporaries do is to make the term meaningless. Instead we should understand it in a useful and significant way. A proof is a communicable symbol of a valid intellectual process. It is part of a communication process by which one person conveys to another his own valid insights and judgments. A proof is not only a proof of something; it is also always a proof for someone. That which is a proof for Karl Rahner is not a proof for John Doe or even Professor Doe. Ultimately a proof never compels. It is true that if a person is committed to logic one can sometimes compel him to admit that he has made a mistake; but this is only because he freely cooperates. His free commitment to being logical outweighs his commitment to his erroneous position. In the last analysis, a proof is a communication and an invitation, not a weapon. Anyone who wants can refuse the invitation, and it is neither possible nor desirable to stop him.

Can the existence of God be proved? Not if one understands "proof" in the modern sense. More than most intellectual processes, the process which concludes to the existence of God depends on implicits which can be symbolized only very inadequately. One has to explicitate the meaning of being, truth, beauty, and goodness, and, as we have seen, this can be done only in an indirect and vague way. Nevertheless, it can be done. The traditional "five ways" of St. Thomas present an outline which has been filled in in modern times by Rahner and others. The heart of the matter is an adequate understanding of symbolizing activity gained from our human experience of our own existence as thinking, desiring creatures.

No proof of a nontrivial truth is ever flawless. As Whitehead said: "If we consider any scheme of philosophic categories as one complex assertion, and apply to it the logician's alternative, true or false, the answer must be that the scheme is false. The same answer must be given to a like question respecting the existing formulated principles of any science." No one had better reason for an awareness of this than Whitehead, who, with Russell, had once launched a magnificent but unsuccessful attempt to reduce mathematics to logic. But nevertheless "the scheme is true with unformulated qualifications, exceptions, limitations, and new interpretations in terms of more general notions." [14] These qualifications are supplied implicitly by the person who uses the scheme even though he is unable to explain what they are. Because our thinking involves implicit knowledge and heuristic anticipations, we are frequently more certain of the conclusion than of the route by which we arrived at it.

In summary, a proof is not a weapon or a machinelike decision-process. It is a helpful communication from one person to another intended to help the second attain the same valid understanding as the first already possesses. It symbolizes the reasoning process of the prover with sufficient completeness to satisfy the standards of both parties. Clearly enough, all this remains true if one of the parties is not an individual but a moral person, like a jury or a scientific community. And, clearly enough, the satisfactory standards for proof will vary wildly from situation to situation. One of the main concerns of a professional knowledge-community is to set such standards for its members.

There is no possibility of positing a set of criteria whose fulfillment guarantee the validity of a proof or explanation. One can indeed formulate maxims which are generally reliable, but there is always a gap between conformity to the maxims and certainty about validity. This gap is filled only by consciousness of personal authenticity on the part of the prover and faith on the part of the auditor.

7. KNOWLEDGE AND FAITH

The fundamental pattern of human knowing which we have outlined applics to both "natural knowledge" and "supernatural faith." In both cases the act of understanding is also a free decision by which one takes a stand toward what is known. One accommodates oneself in a due and ordinate manner to the reality which has already actualized one by a constitutive relationship. In the case of religious faith, the relationship is a personal one with God transcending the texture of the cosmos and demanding a radical personal decision about oneself and the ultimate meaning, the ultimate goal, of one's whole existence. Such a decision engages human freedom more fully than any decision about the created universe, for it concerns itself with the source of our personhood. It affects the relationship which founds our freedom and creativity. Consequently, the freedom of a faith-decision is more obvious than the freedom of an act in which one accepts a scientific theory. Nevertheless, even this act is free and involves values and commitments. One of the merits of Kuhn and Polanyi is to have pointed out that the acceptance of a new scientific theory is a conversion effected by the persuasive power of considerations which cannot be fully specified.

Man's basic constitutive relationship with God, whose term is his primordial heuristic anticipation and desire, is a supernatural one. This is to say, that it is a relation of sonship which goes beyond the relationship required for mere existence and is the result of a special generosity and graciousness on the part of God. Within the horizon thus established, supernatural faith becomes possible. Because man is already oriented toward intimacy with the divine persons, he is able to recognize the meaning of the signs of friendship which God extends to him and to respond to them. Thus, man is endowed from the beginning with a "supernatural existential," that is to say, an element of his primordial desire and implicit knowledge is an orientation toward an intimacy with God which exceeds the relationship required for existence as a rational creature. As we shall see later, this relationship is not a solitary one but an element in a complex of relationships binding all intelligent creatures into a whole centered on the Incarnate Word.

NOTES

1. Aristotle: *Metaphysics*, Book A, Chapter 1.
2. The development beginning here draws heavily upon CORETH, whose book presents a species of "transcendental neo-scholasticism" with which I am generally in sympathy.
3. *Webster's Seventh New Collegiate Dictionary*.
4. LONERGAN, p. 497ff. This does not mean that I agree fully with him in all respects.
5. GILSON, Chapter 1.
6. AQUINAS, Pars I, ques. 13, art. 11, *sed contra*.
7. Thus the Nicene Creed.
8. Compare WHITEHEAD (III), pp. 32ff. Note that the unity and coherence of the universe depend not only on interaction between the parts (efficient causality) but also on formal causality.
9. RAHNER (III), p. 224.
10. RAHNER (III), p. 228.

11. POINCARÉ.

12. Here we see the fundamental ground of the possibility of neurosis. There is a connection between sin and inharmonious psychological development. Obviously this does not mean that a neurotic person is more sinful than anyone else; often enough he is less so than the "normal" persons who force him into or keep him in it. However, I believe that there is always some complicity on his part as well.

13. See LAKATOS for an excellent (and sometimes unwitting) illustration of the impossibility of pinning down the method of even the most "objective" of the sciences, physics.

14. WHITEHEAD (I), p. 13.

6

Creation, Predestination, Spirit

1. CREATION

THE COSMOS IS CONTINGENT. The root of its being, being itself, is not a part of it, does not belong to it, even though it is
within it, sustaining it. This thematic statement is already
implicitly known in the primordial awareness of oneself and
one's world within the horizon of being. Being cannot be
reduced to the level of beings. On the contrary, beings, in-
dividually and collectively, can exist and be intelligible only
within the unlimited horizon which always transcends them.
This transcendence of being with respect to beings implies that
they are not necessary consequences of the inner nature of
being but rather contingent possibilities which have been
actualized. The whole cosmos is therefore a symbol which
emanates from a free decision of God, a decision which could
have been otherwise.

The freedom of this decision must be distinguished from
that freedom which is identical with the divine essence. The
eternal processions which occur within the Trinity (namely,
the generation of the Son and the breathing-forth of the Spirit)
are free because they are totally spontaneous and are determined by or dependent on nothing outside God. The divine

persons constitute themselves in being, and no reason can be assigned for this other than the persons themselves and their decisions (which are identical with the processions). These same acts can also be said to be necessary because they are not contingent. They are not dependent on or conditioned by anything outside God. Thus, in the case of the divine processions, freedom and necessity are identical. The persons and their acts are the ultimate explanation from which all concepts are derived and in terms of which they must be understood.

God's decision to create is free in a somewhat different sense. It is not determined by, but it is dependent upon, the eternal processions. It takes place within the context set by these processions. However, it is not a necessary consequence of them. God does not need creation. His being is already established and fulfilled in the infinite acts which take place within the Trinity. But because He is infinitely good, He freely chooses to share His life with creatures.

This posits something within God, namely, the creative decision in virtue of which He actually is creator. This decision is contingent in the sense that it need not have been and is dependent upon the ontologically prior divine processions.[1] It is not, however, contingent in the sense that it makes God dependent on anything outside Himself. He is creator because He freely chooses to be such.

The decision to create is identical with the very emanation of creation from God. The created universe is the symbol of God precisely as creator. As such it is His fulfillment, or glory, as creator; a glory, however, which adds nothing to Him which was not already there in a higher way in His very essence.

Just as the term "God" refers in a general way to the three divine persons, so the term "creative act" refers in a general way to a complex reality in which each of the persons plays a distinctive role. The Father is the source from which creation proceeds as His symbol. Since it is a symbol of the Father, creation stands in a relationship to the Son who is its center and exemplar. Finally, since the Holy Spirit is the bond between Father and Son, He must play a role in relating the created universe to the Father. Creation makes a difference to each of the divine persons: the Father has created the cosmos and has given his Son to save it; the Son has become a part of it and has suffered and died for it. The Holy Spirit has been poured out upon creatures in such a way that He is now their Spirit, just as he is the Spirit of the Son.

2. PREDESTINATION AND THE HISTORICITY OF GOD

Here we must touch upon a question which has a long and controverted history: predestination. Was the entire history of the universe precontained in the original decision to create? Or is that decision instead a sphere of possibility within which different events can come to be in dependence upon the free decision of creatures? I have already stated implicitly that I consider the second alternative to be correct. The very intelligibility of freedom contains the notion of genuine creativity. When a free being makes a decision, something genuinely new comes into being, something which was not precontained in its causes in a determinate way and which is attributable to the creative activity of him who decides. This genuine though limited share in God's creative power is possible for the creature only because he exists within the sphere of possibility which is the original creative decision of God. So human creativity must be underwritten by God before it can exist, and what man can decide always remains within the limits set by that original decision. There is no doubt, therefore, that nothing can escape the will of God. But part of His will is the freedom of creatures to make decisions for which they and not God are responsible.

I have already indicated that finite causality should be conceived of as analogous to that of the Son in the spiration of the Spirit. On the personal level this causality is free, just as that of the Son is free. The acts of finite agents, especially of free agents, participate in the spiration of the Spirit and by doing so return to God. The Spirit is not an agent within the Trinity, and since this characterizes His personality, the same thing must be true of Him in His relation to the cosmos. Rather, He is the term of acts by which created entities relate themselves to the Father in love. In his *Hymn to the Universe* Teilhard speaks of Power, Word, and Fire.[2] The Holy Spirit is fire, not an agent, but the act of agents which are on fire. The set of possible relationships which He can have to creation determines the possibilities open to finite agents as well as to the Incarnate Son. These possible relationships are real possibilities because of the creative decision in which the three divine persons participated, each in His characteristic way.

Although the original creative decision of God allows for the freedom of creatures, certain things were decided from the very beginning. The human existence of the Son of God is what is most important and valuable in the cosmos, and so this must have been intended by God as a necessary historical event—indeed, as the center of all history, the hinge on which it all turns. What is second in importance about the cosmos is the other persons who come into existence within it in relation to Christ, their brother and exemplar. In the light of the fact that each person is essentially constituted by his unique relationship to God, it seems that His original creative decision must have involved them also and that the existence of each was predetermined from the beginning. The particular course of our lives, the conditions under which we and others live, depends on the innumerable free decisions of countless individuals. Indeed, as we shall see, the fundamental order of the world and the course of evolution is the result of free decision by creatures. But each of the possible cosmic histories contained in God's original creative decision was a particular working-out of the freedom of the same personal beings who were known and loved in their unique individuality from the very beginning.

Jacques Maritain has written of a "sixth way" of establishing the existence of God. It is a development of a basic intuition that it is impossible that a personal being such as I should not have had "an eternal existence in God before receiving a temporal existence in my own nature and my own personality." [3] In a similar way, it is impossible that a personal being such as I should not have been more than a bare possibility (an unlikely one at that) from the very beginning. Each person had to be real in a certain sense from the moment of creation. Otherwise one ends up with irrational chance decisive for personal existence. If the concept of person is indeed the master concept in the order of intelligibility, and personal reality is central in the order of existence, then it is neither intelligible nor possible that material and temporal relationships determine the essential being of persons. That can only be done by the creative decision of God which determines the existence of persons first and of material entities and relationships in dependence upon them.

What the creative decision of God intends is not only a set of persons but a society. Like the divine persons, finite persons could not be themselves unless they were ordered to one another, as well as to God. The being which each of us possesses is a symbol proceeding from the creative decision of God. We are not the whole of creation so we are not an adequate symbol. Nevertheless each of us is, in a certain sense, the whole universe, or perhaps better, a face of the universe. That which we symbolize is one, and in that unity not only we but all other things are precontained. Hence the self-transparency of our being is adapted to and reflects in some way the others who together with us proceed from God, especially the incarnate Word for whose sake the whole exists. Our being is a being-with-others-in-the-world-centered-on-the-incarnate-Word, and our intelligibility is that of a small part of a highly coherent pattern whose central unifying element is the same Word.

Although the original creative decision of God precontains all things which can emerge in history, it precontains them precisely as real possibilities. To become fully actual, they need both the activity of creatures and God's conserving and concurring activity undergirding it. The strong notion of symbolizing activity—which is the fundamental principle of this book—implies that the actual fulfillment of possibilities makes a difference to God. The symbolizer is fulfilled and actualized by the act of bringing forth the symbol. Even though God's essential being is complete without creation, He cannot be a loving creator until creatures emerge from His decision to create. Nor can He engage in dialogue with His adopted sons or respond to their love for Him unless He is genuinely concerned about their salvation and unless His knowledge and love of them is different when they are saved than if they are going to hell. Thus the original underwriting of history by God does not mean that He does not live and act in history, both in a general way by conserving and concurring with all existence,

and in a particular way by signs and miracles and individual
personal responses to those who love Him. God now lives with
men in time. His essential being is eternal and exists in perfect
fulfillment beyond the reach of time and change. But there is
a contingent element in the existence of the divine persons, an
element which is the reality of the divine condescension, the
kenosis (κένωσις [Ph 2:7]), by which the Son has become our
brother, the Father, our father, and the Spirit, our spirit.

Therefore, we must say that even God does not know what
we will do. He knows the possibilities, which exist only because
He has willed them, but which ones will become reality is our
decision. We are creating our own fate and the fate of others.

This is a curious, even paradoxical thing. It is the core of the
mystery of evil. God must put Himself at our disposal. Since
He is intelligent, He can only do this knowingly; He knows
all the possible consequences. But which of these possibilities
will be realized, He does not know until the moment when real
effects emanate from Himself and from creatures united as a
single principle in either love or the perversity of creaturely
sin. Any act of creation is therefore a sort of incarnation and
kenosis of God. The Father must put Himself at the disposal
of others, and in doing so He makes His love subject to abuse
which it is beyond His power to prevent once His original
decision is made. Creation involves risk for God.

It is worth noting that many theologians have refused to
admit the possibility of a real relation of God to the cosmos or
of a contingent and temporal element in His existence. Their
reasoning is that, since He is by nature the infinite fullness of
being God already possesses all that is unqualifiedly good and
cannot change or acquire anything new without admitting imperfection into His nature. This reasoning, however, is based
on a notion of perfection which is unacceptable. A rock endures
by resisting change and ignoring insofar as possible the rest of
the universe. But a man endures by being open to interaction,
sensitive, vulnerable. His stability is a dynamic one which maintains itself by integrating threatening change into the higher-
order pattern of adaptive activity. In the case of God, this kind
of stability is carried to its limit, or rather, beyond all limits.
The divine persons are completely open, totally responsive, and

are constituted by their interaction with one another. Because of this, their union is complete, and the pattern of their common life of love is eternal and beyond the shadow of unsatisfied need, imperfection, or loss. But the very qualities which make the divine persons by nature independent of anything outside themselves also enable them to choose freely to become dependent, in a certain sense, on creation. The Father is able to create other children, the Son is able to become their brother, and the Spirit their spirit. The freely chosen openness of God to creation aims at regaining again a total dynamic stability, invulnerable to change, imperfection, or decay, by bringing creatures into full union with Himself. God is humble and, therefore, capable of entering into time in order to bring the things of time to eternity. Really, of course, His doing so is not debasement or loss. Neither is it gain for Him, merely for us. He already contains all the good of creation within Himself in a higher way, and if we refuse His offer of love, the loss will be ours rather than His.

The notion of being and perfection which would deny God real relationships and involvement in time is really verified only in nonbeing. Nonbeing is invulnerable, unchanging, and cannot lose what it has, namely, nothing. This kind of perfection is what sin aims at, and it therefore appeals to sinful man. As a result philosophical thought which is not guided by revelation tends to arrive at this concept of God.

The concept of divine perfection presented in the scriptures has nothing in common with this. The Old Testament testifies clearly to the transcendence and infinite power and perfection of God, but at the same time it speaks of Him in anthropomorphic and mythic terms. He is vitally concerned with everything He has made; He hears the cry of His people and comes down to them; He makes a covenant with them, cherishes and rejoices in them, grows angry at their misdeeds, and punishes them. Then He repents and restores them to His favor. All these anthropomorphisms are the more striking because of the constantly reiterated assertions of the divine transcendence which are interwoven with them. God is involved with us and is even like us, but in a way which is beyond human comprehension and which excludes all imperfection.[4]

3. MATTER AND SPIRIT

Piaget has argued that thought is action which has been internalized. One could equally say that action is thought which is externalized. During the sensorimotor period the child thinks by manipulating external objects. As he grows older, he learns to behave, i.e., to think, in a way which is more internal and which does not have immediately obvious external effects. Nevertheless, the ability "to think with one's hands" remains an important component of practical activity. At any stage the symbol which makes explicit man's subjective knowledge and desire is the total pattern of nerve excitations, muscle contractions, glandular secretions, and chemical processes which constitute his total behavior. Thus symbolizing activity is the integrated totality which includes understanding, love, motor action, perception, bodily desire, even metabolic and reproductive activity insofar as they are controlled by the organismic regulation of the personal center. In the lower organisms, which lack a highly differentiated structure, behavior is not highly differentiated; in man, whose structure *is* highly differentiated, behavior is highly differentiated. The component which goes on in the brain, the most complex and reactive substructure of the organism, is the most rapid, complex, and expressive of all.

The conception is relevant to developments in contemporary psychology. There is evidence that the person who is cured of neurotic difficulties or who grows in his personal integration is the one who can understand and express himself both in terms of feeling and in terms of language. It is a matter of "bodily felt meanings" as well as of conceptual meanings.[5]

The realization that rational thought is a particular kind of bodily expression leads to a better understanding of the relationship between the "spiritual" and the "material" elements in man. Briefly, the human spirit is an organizing dynamism which expresses itself in matter. The body is the symbol thus constituted. Let us discuss this in more detail.

There is really not much problem about the *existence* (as opposed to the nature) of the spiritual. The human subject, whom each of us knows so intimately though confusedly, has remarkable characteristics which warrant the use of a special term. This subject is spiritual by definition.[6] Neither is there much problem about the existence of the material. We find in our ordinary experience entities which are simpler and less significant than ourselves, such as water, rocks, air, molecules and atoms of all sorts, etc. These entities exist, but they give no evidence of being aware of themselves or of being free. They do not possess themselves, do not exist for themselves in the same full way in which men do. In other words, they are not spiritual but material; they have a limited existence which is closed to itself and which lacks the self-transparency of knowledge and love. The extraordinary thing is that material entities are capable of becoming parts of us. Our experience of our own lives gives us every reason to believe that our bodies are indeed a part of us and that these bodies are made up of atoms and molecules which we assimilate from our material environment.[7] What then is the relation between the spiritual and the material and what is the ultimate nature of both?

We have already rejected the reductionistic explanation that the human person can simply be reduced to the material. We have also rejected the dualistic explanation that the human subject and matter are simply disparate and are related to one another in a rather tenuous and almost totally mysterious fashion. There is a third untenable explanation which has already been implicitly rejected but which I want to discuss explicitly now. This I shall call "emergenism." It is the application to man of a theory which is essentially correct in regard to the lower animals. Because man is an animal, it is partially correct in his case as well, but because he is more than an animal, it not only is incomplete but constitutes an implicit denial of what is most important about him.

One who holds emergenism can agree with everything in the first part of this book. He regards the universe as a system ordered hierarchically in such a way that the entities on each level have a depth of being and intelligibility which cannot be fully explained in terms of the lower-level systems which they integrate into their unity. During the course of evolution more and more complex structures have emerged from the potency of matter to bring the cosmos to higher and higher levels of being, truth, beauty, and goodness and in man the cosmos has attained its highest level to date. Man indeed surpasses all other entities (except the universe itself), but emergenism asserts that there is nothing in him which transcends matter. He is the realization of a potentiality which was within matter from the beginning and which does not require any kind of special causality to become actual. The gap between man and the chimpanzee is not radically different in kind from the gap between chimpanzee and flower or between flower and rock.

To hold this is to undervalue radically the uniqueness of man and the worth of his intelligence and freedom, his love of beauty, moral goodness, and truth. Much better than crass reductionism, emergenism is, nevertheless, an attenuated form of it. The characteristic abilities of man are not a mere improvement of abilities which the animals already possess. Animals are able in some measure to understand the concrete embodied patterns present in nature, and to construct and use symbols. But unlike man they never grasp the pattern as such or understand the symbol as such. They do not situate the actual within a wider framework of possibilities stemming from their own implicit understanding of being. This can be done only by a being whose essence is implicit knowledge and desire of the ground of being. It is because we are aware of ourselves as being-in-the-world capable of expressing our implicit knowledge and desire in symbol that we can understand the meaning of being, of knowledge, of desire, of symbol.

One of the major surprises to students of evolution during recent decades was the discovery that humanity and rudimentary culture and speech antedate the development of the large brain. It is generally agreed that the Australopithecines, who inhabited south and east Africa a million years ago, were human. Yet their cranial capacity was no larger than that of the modern ape.[8] The reason, obvious in retrospect, is that there is no point in having a large brain unless one has some use for it. First came the species, man, which wanted to understand and communicate intelligibility. Then came the organs required to do so efficiently. A large brain has no survival value for an animal which does not possess the heuristic anticipations needed to make use of it.

Because they are not directly ordered or related to the ground of being by implicit knowledge and desire, brute animals cannot be subjects in the same sense in which men are. They are not ultimate and creative sources of symbolic structures. This is to say that they cannot understand, love, or act in any ultimate sense. The ability which they have to understand and desire is in them in somewhat the same way in which the ability to calculate is in a computer. The computer performs operations which are isomorphic to some of the formal operations of intelligence but which lack the creative awareness, the tacit dimension, of intelligence which turns formal operations into truly intellectual understanding. Human intelligence can limit itself and project an aspect of itself into a computer, but a computer cannot remove its limitations and understand. Similarly, there is an animal level within man which, as B. F. Skinner and the behavioral psychologists have shown, contributes substantially to human behavior. But man transcends the limitations of the animals.

In other words, there is a certain dimension or aspect in which man is infinite, without limits. This dimension is in him because of his direct personal relationship to the unlimited ground of being. Being itself is infinite; when it turns in love to the cosmos and speaks to it a personal word, the symbolic term of this act which arises in the cosmos is also unlimited along the dimensions of love and personal knowledge. Otherwise it would be incapable of existing in a personal relationship with its unlimited ground. And conversely, without a direct relationship to that which is unlimited, man himself could not be so and could not understand or love.

To know is to situate the beings which are known within the unlimited horizon of being which is already present within the knower, the horizon which is identical with his primordial heuristic anticipations and desires. A being which has this unlimited desire within him must perform acts of knowledge and love. A being which has not cannot do so; it lacks the desire, and its horizon is too small to admit the being of beings. It can understand beings in terms other than being, as animals do. It can observe patterns of phenomena in space and come to know phenomenal objects; it can observe patterns of phenomena in time and come to know regular connections; it can observe the positive reinforcement supplied by certain phenomenal objects and know their suitability for its own ends. But it cannot know being, causality, goodness, or beauty. These intelligibilities only exist within an unlimited horizon.

Animals, like computers, lack creativity. They are not ultimate sources of symbolizing activity. They may be pushed or pulled into situations which establish within them the patterns needed to secure positive reinforcement from the environment, but these patterns are on the same level as they are and the goods by which they are reinforced. Creativity and freedom are one and the same. Man has freedom because within a horizon without limits he can make choices in which he specifies and symbolizes in a way for which he is personally and ultimately responsible.

Whether man is unique is a fundamental option, and the answer given depends on one's whole outlook and personal philosophy of life. In my view, a negative answer to the question of the uniqueness of man makes impossible a genuinely moral condemnation of the gas chambers of Nazi Germany and other horrors of this century. It also ignores the intelligibility of symbolizing activity given in our subjective consciousness.

How does the spirit which is the partial cause of the body arise? It is clear that even the spiritual human person is caused, at least in a partial sense, by the cosmos as a whole. We can call this person spiritual and thus emphasize his distinction from brute matter, but he too is a part of the ordered cosmos. He is composed of matter, i.e., his subsystems are material, and without matter he would have no field of action, no means of self-realization. He is therefore unintelligible and impossible except in relation to matter, and must be partially constituted by his relationship to matter.

However, as we have seen, what we might call the "personal center" or the "spirit" of man is constituted by a direct relationship to the ground of being, that is, to God the creator. A man is therefore a being who transcends the cosmos even while he is a part of it. He is a hierarchically ordered structure who develops out of and in dependence upon the material universe. But at the same time the concrete unity which is an individual man could never arise except as a limited symbol of the infinite ground of all beings, the Father, who constitutes him as the term of a unique personal relationship which is patterned on the relationship of Father and Son. Personal beings are the terms of personal relationships to God. In this resides the unique and irreplaceable value of each individual man, a value which is incompatible with emergenism, even a theistic version of it.

The human spirit is not a complete nonmaterial entity. Rather, it is the Teilhardian "inside" of a man, an aspect of the concrete being which is the term of a direct relationship with God. A person is a "face" of the cosmos which arises when God addresses it. Each of us contains the whole within him, even while we are a part of the whole, because of the fact that our self-expression is not equal to the whole but is limited, at least during this phase of our existence, to that limited substructure of the cosmos which we call our body.[9]

Therefore I am denying the dualistic assertion that man is composed of body and soul, as of two distinct entities. I am also denying the monistic assertion that man is simply material. I am also denying the Aristotelian assertion that man is composed of two distinct principles, matter and form, where the form is evoked from the potency of the matter. Man is a concrete unity which has two aspects, a spiritual aspect arising from the divine causality and a material, or better, a cosmic, aspect which arises from the causality of the cosmos. The traditional doctrine of "the direct creation of the soul" is the expression of an essential truth about man.[10]

The concrete unity of the individual man, which includes the spiritual personal center and its bodily expression within a single whole, arises at a definite moment of time from the joint action of God and the cosmos. The part of the cosmos in this is instrumental. Its existence and its powers have been prepared in advance so that at the right moment it may contribute to the production of an effect which is beyond its unaided power. Thus the action of the cosmos transcends its nature in somewhat the same way as the intelligence and desire of man transcend his nature. All finite beings can act in a way which is beyond them and which must be underwritten by God in order that it may be possible. Though it acts beyond itself in causing man, the action of the cosmos is necessary. God's creative relationship terminates not in an independent pure spirit but in a center of activity which is precisely an activity in matter, a spiritual being but not an independent nonmaterial entity.

We do not know at what precise moment in its development the human body becomes human in the proper sense of the term. We can say abstractly that it is at that moment when it becomes a symbol of the implicit knowledge and desire which is the essence of spirituality. Before that it is merely highly organized matter having the same kind of unity as the body of a higher animal such as a chimpanzee. It becomes properly human only when it becomes explicit, thematic knowledge and desire.

It is conceivable that this takes place very early in development and that there is some type of very vaguely thematized human consciousness even in the womb. It is also conceivable that this does not take place until after birth when social interactions begin. We usually think of hominization as the result of the sheerly physical processes of assimilation and growth. But it might be more plausible to think that it takes place only when the environment demands a response on the human level from an organism which has substructures capable of being organized into a human whole. In other words, hominization would be a stage in intellectual growth. Such growth usually depends not only on the maturation of physical structures but also on interaction with a suitable environment, especially a social environment in which specifically human interactions are common. In this hypothesis hominization would be akin to, and perhaps identical with, the attainment of a Piagetian stage of organization. It would, however, be different from the attainment of most such stages in that it is a substantial rather than an accidental change and one which takes place at a definite instant of time.

Analogously to this we can note that the hominization of the first man need not be attributed to superior genetic endowment alone. Doubtless his genetic endowment was superior to the average of his nonhuman associates, but part of the superiority which enabled him to become human may have been learned behavior patterns which set the stage for further ad-

vance to the level of truly human behavior.[11] Of course, these
behavior patterns would have been just as real and physical as
chromosomes.

NOTES

1. Here we first encounter something like time in the life of God.
God creates because He already is good, loving, and powerful in His
triune way. Therefore creation comes "after" the self-constitution of the
three persons. In contrast, the Son is "ontologically consequent" to the
Father, but not "after" Him since the Father Himself is constituted by
His generation of the Son.
2. TEILHARD (III), p. 21.
3. MARITAIN, p. 76.
4. Why was the prima facie impression given by Scripture dismissed
for so long by systematic theology? Perhaps it was a combination of
accidental factors in the history of philosophy (the Greek tendency to
view perfection as something static) along with fundamentalistic
exegesis of certain isolated scriptural passages. In any event the work
of Whitehead and process theologians (even if one does not agree with
them entirely) has caused many people to take another look at the
question and to adopt a different viewpoint. I believe that the goals of
the process theologians can be better attained by making use of
Rahner's concept of symbolizing activity than by use of Whitehead's
categories.
5. See GENDLIN.
6. This is to say that spirituality and subjective self-awareness are
closely related. The use of the special term "spiritual" emphasizes the
ontological distinctiveness and centrality of what materialists claim is
only an epiphenomenon. See RAHNER (IV), p. 162.
7. Though this is clear enough, it is not as clear as the basic fact
that we are "spiritual" beings.
8. DOBZHANSKY, p. 200.
9. After death the situation is different. I am attracted by Rahner's
suggestion that after death the human spirit has a "pancosmic" relation
to matter. See RAHNER (I).
10. The form in which this doctrine is expressed is often antiquated.
But simply discarding it results in irreparable loss. It happens here as
it does so often that rejection of tradition is not the imaginative and
creative thing to do. What is needed is the more difficult task of
deeper understanding and more adequate expression.
11. GRELOT, 468.

/

7

The Problem of Evil

1. EVIL: PROBLEM, MYSTERY, AND TESTING

THE PROBLEM OF EVIL is probably the chief obstacle men encounter in their search for understanding of the world. "Every human order is a community in the face of death," [1] and theodicy, the attempt to integrate "anomic" phenomena like successful injustice, suffering, and death within the world view of the community, is one of the key issues to which every religion must address itself. This is especially true of Biblical religion with its belief in a God who is at once omnipotent and loving. As the writings of Camus illustrate, the breakdown of the plausibility of the Christian explanation of evil is at the root of much modern atheism.

To me it seems an obvious, though often overlooked, fact that the problem of evil is partially a problem about the structure of the cosmos. Hence it is a problem which must be faced in this book. I shall make no attempt, however, to deal with it in all its aspects and ramifications; that would be a vast undertaking indeed. Instead I treat it in terms of the viewpoint of this book, that is, mainly in its relationship to cosmic structure.

To clarify the meaning of this it may be helpful to distinguish between the problem of evil and the mystery of evil. All true

143

mysteries ultimately derive their character from The Mystery, which is God. A mystery is a luminous darkness in which we contact a reality too great for our symbolizing power; therefore we cannot know it with the clarity of explicit knowledge. We will understand mysteries only when we understand the divine processions and are thus enabled to grasp clearly the nature of symbolizing activity. Hence evil, whose core is defective symbolizing activity, cannot be fully understood until we understand God. It is an aspect of the mystery of freedom.

The problem of evil, on the other hand, can be understood as something more superficial than the mystery of evil. It is concerned with our attempt to construct a coherent theory of the cosmos and to relate this to our understanding of God. Its solution is therefore attained when we have put evil into a coherent relationship with the mystery of freedom and the mystery of God. In this sense the problem of evil admits of a solution. But solutions, being conceptual things, are never absolute truth. They are always more or less adequate, more or less marred by sin and imperfection. The solution we have is always an approximation, a term in a sequence which approaches the ideal solution asymptotically. When new aspects of the inexhaustible mystery of being appear to us, we have to expand and revise our old conceptual schemes in order to accommodate and clarify our new grasp of reality.

This seems to be what is needed now. The traditional Christian understanding of evil in terms of the doctrine of Original Sin has collapsed under the impact of new knowledge about the cosmos. I believe that a satisfactory solution—that is, a conceptual scheme which adequately (for the time at least) re-expresses the truth contained in the old view and integrates it with the new knowledge—grows naturally out of the framework which has been developed in the preceding chapters.

I want to emphasize that a conceptual scheme cannot be identified with the personal wisdom by which a man integrates his own experience of evil into his mind and heart. Such a scheme may be a help. Without it one can be tempted to seize upon faulty conceptualizations, e.g., that the world or oneself is absurd, that God is a sadist, that good and evil are identical or else illusory, etc. But even if one has a coherent conceptualization in one's mind, when the terrible force of real evil falls upon one, the matter assumes a new dimension. A conceptual scheme may be coherent, but is it true? And even if one secretly inclines to believe that it is, does one wish to admit this and thus cut off the easier outs of despair or hatred? Concepts seem awfully pale in the fierce light of agony. They will only have staying power if they are transfigured by the power of love which raises them to another dimension where they become essentially subordinate elements. In the end the existential answer to the problem of evil is the cross of Christ.

Therefore a conceptualization of the nature of evil is an empty framework which needs to be filled by personal spiritual experience of the truth of the Christian revelation of the love of God in Christ Jesus. And once it has been so filled one realizes that the framework is far less important than the revealed truth which transfigures it.

How probable is the conceptual scheme to be developed here? In the preceding chapters I attempted to integrate ideas which have already been developed, at least in part, by eminent modern thinkers. The key idea of this chapter may be more controversial because it is not "in the air" right now, at least not in the air breathed by people over thirty. It stands in opposition to some deeply seated heuristic anticipations of the modern period.[2] As far as my personal judgment is concerned, this does not make it any less convincing; that is to say that for me it has high verisimilitude. But the judgment that it is true depends, even for me, not only on individual judgment but also on social reinforcement. That is to say, in Piaget's terminology, that complex truths are attained by a process of equilibration. The scheme seems to grow naturally and spontaneously out of the data of revelation and personal experience which I hold as certain, but there is also an element of self-commitment to hypothesis involved. Hence it seems that the truth-value of the following considerations will become known with certainty, even to me, only through a long process of sifting and weighing. And since the scheme is related to the Biblical revelation about evil, the weighing will have to be religious and theological as well as philosophical.

2. THE GROUND OF THE COSMOS

We begin by trying to understand the fundamental dynamism expressing itself in the development of the cosmos. We note first of all that it is the total cosmos which develops. It is one system with a definite unity, not just a collection of haphazardly interacting entities. We have already seen that, when we reflect on them philosophically, even the laws of physics reveal this. It is not coincidence that Newton's gravitational formula is obeyed closely by matter in all parts of the universe. The Pauli exclusion principle implies that all particles of a single type are excitations of a single system which is co-extensive with the universe. Physicists have not yet succeeded in demonstrating that all elementary particles are excitations of a single field, but many of the greatest (including Einstein and Heisenberg) have had faith enough in this idea to exert strenuous efforts toward making it explicit and rigorous. The whole drive of theoretical physics is toward a unified conception of matter which will reveal particular pieces of it as subsystems of a single all-embracing system. This heuristic anticipation is, I believe, based on a valid implicit understanding that the entire cosmos is one, that in some sense, difficult to specify, it is one being.

When we come to biology and the theory of evolution, we find that biologists insist that it is not individual animals but populations of animals which evolve; in fact, it is the entire ecosystem which evolves. The development of a single species is determined by the development of the other species with which it interacts and by the physical environment in which they all live. But even this perspective is too narrow. For a full understanding of the dynamics of evolution one must take into account the constraints set by physics and chemistry and by the geological evolution of the earth and of the solar system of which it is a part. The complex molecules presupposed for the formation of living beings may have originated in space even before the solar system was formed. Moreover, the nebulae from which solar systems and galaxies condense are the result of the formation of the elementary particles in the early moments of the universe (or at least of this phase of the universe's history). Thus in the end the concept of evolution expands to embrace the whole cosmos throughout its entire extent in time as well as space. There is no dividing the process into neatly separated parts or phases.

This is not a surprising result. Thinkers of all sorts have sensed that the cosmos is one and have expressed this intuition in a variety of ways. To me this basic insight seems certain, albeit mysterious and difficult to specify unambiguously. And just as the cosmos is one, so is its evolution. The process is not a mere sequence of events following one another without any particular internal necessity. It is a genuine development, an unfolding of a basic potentiality which existed from the very beginning. It is akin to the development of a human individual in the gradual unfolding of his unique personal potentialities. In other words, the process is a natural one; it has its own inner coherence and is complete in its own order. Although it can only come to be within the sphere of possibility which is God's creative act, once that act is placed, the rest follows naturally according to its own inner dynamism. God does not first create one state of affairs and then an instant later create another which is connected with the former only in terms of formal similarity. In other words, the dynamism of evolution is not similar to that of a movie in which one picture succeeds another and is so related to it that the latter one seems to emerge from the former even though in actual fact all the pictures are created by the hidden camera. Rather, the existing past symbolizes and expresses itself in another increment of reality which is the objective present moment. To deny this is to fall into a kind of occasionalism which, though it is logically coherent, runs severely counter to our modern apprehension of nature as an autonomous whole.

If causality is to be preserved, such a process of development requires an initial potency which precontains everything which later unfolds into full actuality. Furthermore, this initial potency cannot be a passive one which requires actualization by another. Such an other would be an agent outside the cosmos, that is, God; and once again we would have God doing everything and created beings playing the role of puppets. Rather, the potency which unfolds in the course of evolution is an active potency, a power of action which can fulfill itself by bringing forth a symbol which is at once identical with, and distinct from, itself.

What can we say about the nature of this cosmic ground? It is clear that in terms of being and value it exists on a level at least as high as that of the animals which emerge from it, that is, the level of sentience. Is it plausible to suppose that it possesses no greater excellence than that? It would not be logically contradictory for us to say so since we have argued at length that the characteristic excellence of man is the result of his direct personal relationship with God, a relationship which transcends the texture of the cosmos. Nevertheless such a statement seems implausible to me. The ground of the cosmos is not a short-lived entity like an animal. It is the source of ten billion years of cosmic history and is responsible for the sublime beauty and diversity of the material universe. All the artifacts of man, all his loftiest scientific theories, fall far short of the reality which has unfolded from this ground. Furthermore, even man himself must be regarded as partly the creature of the cosmic ground. We have seen that evolution is a teleological process which aimed from the beginning at the emergence of man, and thus the purposiveness embedded in the cosmic ground extends as far as the emergence of freedom and intelligence.

The physicist Richard Feynman introduced his brilliant series of popular lectures on the character of physical law with the following remark apropos of Newton's theory of gravity: "This law has been called 'the greatest generalization achieved by the human mind,' and you can guess already from my introduction that I am interested not so much in the human mind as in the marvel of a nature which can obey such an elegant and simple law as this law of gravitation. Therefore our main concentration will not be on how clever we are to have found it all out, but on how clever nature is to pay attention to it." [3] I believe that Feynman is giving indirect expression here to a feeling which I share. Though man himself is the greatest wonder in the visible universe, none of his works, not even his intellectual works, can equal the beauty and sublimity of the material universe. When one considers not only the "elegant and simple" laws of physics but also the biochemical intricacies which emerge from them and which ultimately give rise to the neural circuitry of the human brain, it becomes difficult, at least for me, to imagine that the root of the symbolizing process which produced them is anything less than the spiritual dynamism of intelligence.

In other words, at the root of cosmic process is an intelligent being or beings whose symbolizing activity produces the material structures of the universe. The arguments I have given so far do not establish this conclusion with the solidity I should wish. They are based on too many implicits which I am unable to specify. However there remains another argument which is stronger: it is based on the palpable fact that evil exists in the world.

3. DUALISM AND THE PROBLEM OF EVIL

The classical definition of evil is that it is the absence of some good which "ought" to be present in the being or situation which is evil. Evil as such, therefore, is not a being but a privation, a defect of a being which is fundamentally good. The perspective of the present work enables us to be a little more specific. Evil is a defect of some symbolizing activity as well as the defect which it causes in the symbol produced and the lack of fulfillment which results in the being doing the symbolizing.

The ground of all beings, being itself, cannot be evil. God is the source of all intelligibility, the horizon within which everything else must be judged. Since He is the source and the norm of all being, there is no possibility of His being judged, or of His actually being, defective. Being itself is an infinite plenitude which excludes imperfection. Neither can His symbolizing activity be defective. He has nothing to express save Himself, and this is wholly good. Therefore evil must have originated with creatures.

The world as it came from the hands of God had to be wholly good, though limited. It must, therefore, have become evil in the course of development, when defective symbolizing

activity created defective symbolic structures and failed to fulfill and complete the original being with which it was endowed.
A being which is not creative cannot be the origin of evil.
If its primordial being, or first act, is defective, this is the result
of the defect of the symbolizing activity which established it.
If its symbolizing activity, or second act, is defective, this results either from its previously defective being or from a
deficiency in the activity of the being which is now empower-
ing it to act. Therefore, only intelligent creatures can originate
evil, i.e., commit sin. We must conclude then that at some
point in history an intelligent being, or beings, which was good
and without defect freely chose evil and so failed to develop in
the way it should have.

Because it is limited, an intelligent creature without defect
can engage in defective symbolizing activity. Sin is a free refusal to develop in the way called for by one's own subjective
being. It turns guiltless lack of perfection into freely chosen
and desired lack of perfection, and thus orients one away from
God, being itself, toward nonbeing, which is death. It is a
refusal of life, an act of hatred for oneself insofar as one is
ordered to ultimate being and value, and an act of love for oneself insofar as one has not yet achieved it. The being of an
intelligent creature is a being ahead of itself toward God. If
that which lies ahead is refused, being is refused, and death
and nonbeing are chosen.

It is clear that man is a source of evil in the world. But he
cannot be the only source. Although he sometimes freely
chooses evil, it is nearly always because he is tried and tempted
by a world which is out of joint, which makes it difficult for
him to act as he should. Furthermore, even if man had never
sinned and had always acted for the good to the full extent of
his powers, he would still suffer. The whole system of the
world is such that physical evil is inevitable.

This physical evil which is part of the present world order
is evil in a strong sense of the word. It is contrary to the good
of man, the final end of the cosmos. This is true because it
afflicts him physically and psychologically, but much more so
because it makes sin almost inevitable. The fear of death is the
root of an anxiety which men attempt to assuage by seeking

the anodyne of pleasure, by acquiring power, or by creating false systems of meaning. Furthermore, it makes it difficult for men to perceive the goodness and love of the creator in an order of things which will ultimately crush him.

The whole process of evolution has aimed from the very beginning at the production of man and was designed for his good. But it has fallen short of achieving that good in full measure. If the structure which evolution has produced is partially evil, then evolution itself is also partially evil. Therefore the cosmic ground, like man himself, is infected with sin, a freely chosen lack of the fullness of being and value which was destined for it, and through it for man, by God. Consequently this ground is distinct from God, the fullness of being, intelligibility, and value, in whom there is no shadow of imperfection.

If the cosmic ground is involved in sin and evil, are we to think of it as a single being which is sinful and imperfect, or should we think of it dualistically, in terms of two irreconcilable forces, one of light and the other of darkness? Without Christian revelation we could never be sure. But the cosmogonies of ancient cultures were as a rule dualistic in character. They pictured the foundation of the world as a struggle between the gods and their enemies. Among the Babylonians an annual ritual recalled the victory of Marduk over the marine monster Tiamat from whose dismembered body he created the cosmos. Similar ceremonials were held among the Hittites, the Egyptians, and at Ras Shamra.[4] The Old Testament recalls the ancient legend of Yahweh's victory over the powers of chaos (Ps 74:13–15; 89:10–11; Is 27:1; 51:9; Jb 9:13; 26:12; 38:8–11). These ancient myths are naïve expressions of man's implicit awareness of the nature of good and evil. Good and evil are not like black and white pigments which can be mixed in varying proportions to form a continuous spectrum of grays. They are dynamic powers which struggle for mastery within the cosmos and within man. During the brief period between its birth and death, they can exist together within the same human spirit, but in the end they are irreconcilable. Every spiritual being must sooner or later choose whether he will advance toward God or turn away from Him. The Powers which ground the cosmos are not partly good and partly evil. Some are good and others are evil, both in an integral way. The correctness of this intuition is confirmed by the New Testament in which the work of Christ is presented as the salvation of man from the power of the "Adversary," "the Prince of this World," who had held him in thrall. This dualistic perspective, which will be discussed at length in the following sections, implies that the structures of the cosmos are grounded in two spiritual forces alien to one another, one of good and the other of evil. The development of the cosmos is not a totally integrated process but a struggle in which, just as in human life, growth and progress win a precarious victory over evil at high cost.

Thus the unity of the cosmos is grounded in a society in conflict with itself rather than in a single personality. In some respects, then, the material structures of the cosmos are more like a partially fragmented culture than a human body or the structure of a single human mind. Culture is the common creation of many human minds and, thus, has a kind of objectivity and inner inconsistency which transcends the objectivity and inconsistency of one individual's private mental life. This is part of the explanation of why the human person's bodily expression is not totally at his own disposal. Undoubtedly there is a unique personal quality to our bodies, but the possibilities of expression open to us are limited by the societal expression which is the material structure of the cosmos.

The two intentions which underlie the cosmos are embodied concretely in a single structure. The angelic will aims at absolute good, but it has had to participate in the construction of a world marred by evil. It was better that evolution proceed at the cost of suffering and death than that it not proceed at all. In the morally ambiguous structure which has resulted, we can read two messages: on the one hand all that points and leads to Him who is beyond the cosmos, whose names are Being, Truth, Goodness, Beauty, Love, Joy. But mingled with this message comes another: absurdity, pride, hatred, despair; a reflection of the great but ultimately foolish intelligences which have chosen slavery to their own limited natures rather than the freedom of the sons of God. Not many men would have been able to untangle the two adequately without the aid of explicit revelation.

4. THE TRADITIONAL DOCTRINE OF ORIGINAL SIN

Christianity attributes the existence of evil in life to Original Sin. We can use the term "Original Sin" in two senses: Original Sin *originatum* denotes the present sinful state of man in which he is inclined to moral evil and subject to suffering and death; Original Sin *originans* denotes the historical act by which intelligent creatures introduced evil into human history and created the situation of involvement with evil from which we suffer today. Traditionally Original Sin *originans* is attributed to Adam and Eve, the Biblical figures of Genesis 2 and 3, who are identified there as the first two human beings and the ancestors of all men now living. I shall argue that Adam and Eve are mythic figures who never existed and that in fact Original Sin *originans* was the act of some of the Cosmic Powers who are the ground of the cosmos and who have involved not only man but the whole cosmos in sin and evil.

The traditional Christian teaching on the nature of Original Sin *originatum* was that it consisted in the absence of supernatural gifts (such as sanctifying grace, faith, hope and charity, and the gifts of the Holy Spirit) and also of certain "preternatural" gifts, among which were immortality, impassibility, and integrity. Immortality and impassibility conferred immunity from bodily imperfection, suffering, and death. Integrity gave man the power to dominate his nature in such a way that it never impeded, but always responded fully to, personal free decisions.

This seems to be an accurate diagnosis of what is wrong with the human condition. Our most terrible (though not the most obvious) problem is our lovelessness, our lack of love for God, first of all, and for our neighbor. But this problem is partly caused by the physical evils which we suffer, the greatest of which is death. From death comes anxiety, the fertile seedbed of sin. Furthermore, our free decisions cannot dominate our natures. Even the best of intentions can be undermined by passion. Man is fickle and inconstant; at best he takes two steps forward and one back. In other words, Original Sin *originatum* is precisely the evil which "the problem of evil" is all about. In the Christian perspective it is seen as the result of sin and, in the traditional teaching which held sway before Darwin, of the sin of our first parents, Adam and Eve.

The authors of the traditional teaching saw correctly that the preternatural gifts were in some sense miraculous. The order of nature, as we know it, does not produce them and is even opposed to them. Hence they concluded that God gave them as special gifts of grace.

There was always a certain awkwardness in the teaching about the preternatural gifts. In what sense is their absence a privation? Does man have some sort of claim to them? Or are they purely gratuitous? Either answer leads to difficulties. Men have always sensed that temptation, suffering, and death are somehow wrong and incongruous. This intuition is so strong in some men that it drives them into atheism. Thus in Dostoevski's *The Brothers Karamazov* Ivan refuses to accept a God who would create a universe in which innocent children suffer. Dr. Rieux, the sensitive and noble humanist of Camus' *The Plague,* feels the same way. And even Job might be tempted to draw the same conclusion if he lived in our century. The Christian too feels the force of this insight into the wrongness of suffering and, in fact, goes beyond it in his understanding that the existence of scandal and temptation is even worse. Hence he would like to draw the conclusion that man is entitled to impassibility and integrity and that God's original creative decision must have intended that he have them.

But this conclusion is blocked by the evident fact that the order of the world makes no provision for such gifts and, in fact, seems to rule them out. Since God is the author of nature, this is embarrassing. Man is the final end, the *raison d'être* of the universe. Why then did God not adapt it better to his needs? The uneasy solution to this dilemma was to say that, though these gifts are not called for by the order of nature, God nevertheless intended to bestow them on man in a quasi-miraculous way. It is only because of man's sin that He does not. Thus the responsibility for the present situation is man's.

But this solution is beset with difficulties. First of all, there is the intrinsic one consisting in the fact that the gifts are withheld from newborn children who have never committed personal sin. And even if they had, why should God not give them a second chance? Christ, God incarnate, set a standard of forgiveness which we struggle in vain to attain, but even we would find it rather easy to forgive ordinary people their sins and restore the gifts they need to them. The conceptual twists and turns theologians have made to escape this difficulty are marvelous to behold!

Another difficulty is quite acute today, though it was not much of a problem before the rise of science. As long as nature's operations were semi-mysteries, it was easy to blur the distinction between what was opposed to its order and what was not. Formerly one could view the preternatural gifts as modifications of the natural order which were not fully miraculous. Their existence would simply mean that one arbitrary operation of natural entities would be replaced by another. Because the connections between different properties of the world were hidden, it was not clear that one change ought to require others also. But today we know that nature is a fantastically tight-knit web. A change in the fall of a stone here on earth would shatter the galaxies. The preternatural gifts would require either an endless series of miracles or the existence in man of a hidden power to dominate nature which could be actualized by grace. The first alternative makes it difficult to understand why God ever created the present order of nature in the first place. The second alternative is attractive in many respects; in fact the writer himself adopted it for a time.[5] However, it faces the difficulty that there is little evidence for it, and so it is vulnerable to the perils of Ockham's razor.

I want to discuss now some of the incongruities entailed by the belief that God once intended to insert the preternatural gifts into the present order of nature. The same discussion will also display further evidence in favor of the position taken in the preceding section, namely, that evil is deeply rooted in the present world order and that in fact it affects even structures below the human level which antedate man's appearance on this planet.

First of all, the insertion of the preternatural gifts into the present order of the world would require the suspension of the laws of physics and chemistry. This is so obvious that nothing more need be said. It would also entail short-circuiting the very process of evolution which produced man in the first place. The occurrence of evolution depends upon random or near-random mutations of genetic material and natural selection. But mutations are more often harmful than helpful and, as a result, defects of all sorts are inevitably produced in organisms. Natural selection is almost synonymous with suffering and death. It occurs because the less adapted fail in the ceaseless competition for food and mates or are devoured by predators, parasites, and microbes. And in the end even the healthiest must die in order to make room for another generation.

It should be noted that evolution continued by the same means for a million years after man appeared on earth. The Australopithecines, who lived in south and east Africa some million years ago, are generally classified as human. Nevertheless their cranial capacity was no larger than that of the modern ape. *Homo erectus*, an unquestionably human predecessor of modern man, had a cranial capacity about midway between that of the Australopithecines and modern man.[6] Brain size is not an unequivocal index of intelligence, but it is clear that this increase in brain size over the past million years was roughly proportional to and indispensable for the growth of human intelligence. If Adam and Eve and their descendants had been exempted from suffering and death, Jesus Christ would, naturally speaking, have been incapable of uttering His words of power, His disciples would have been incapable of writing them, and we would have been incapable of reading or understanding them.

5. CONCUPISCENCE AND AGGRESSION

Man's moral failings are also bound up with his evolutionary background and with general characteristics of the whole structure of the universe. Our situation is a "situation of sin" which often solicits us to that which our better thoughts reprobate. This is in large part the result of the difficulty we have in controlling instincts and behavior patterns which were created during the course of evolution and which are no longer very appropriate for a rational being, especially in modern circumstances.

I have already mentioned that the traditional theory of Original Sin attributed man's moral failings partly to a lack of "integrity," that is, lack of the ability to order animal passions and impulses by reason. As a result the spiritual will finds itself with a bias in favor of what is sensibly good, a bias which often results in decisions against the better judgment of conscience. Consequently man has an inclination toward sin, at least in the minimal sense that he is prejudiced in favor of sensible goods which can just as easily be humanly and morally inappropriate as appropriate. This inclination of man is often called "concupiscence."

St. Thomas and other Scholastic philosophers called the powers of man by which he is inclined toward sensible goods the "sense appetites." [7] They distinguished two different sense appetites, the "concupiscible" and the "irascible." The concupiscible sense appetite has as its proper object the sensible good simply as satisfying and fulfilling. Thus man finds satisfaction in food, sex, the gregariousness of the herd and, in general, everything which can be known by the senses as contributing to bodily well-being. The basic acts, or passions, of the concupiscible appetite are love, desire, and pleasure, and their opposites hate, aversion, and pain.

The irascible sense appetite has as its object sensible good which is difficult and arduous to obtain. The passions of this appetite are hope and despair, daring and fear, and the passion which has no opposite, anger. In one way the irascible appetite is auxiliary to the concupiscible. We must first desire something before we can want to overcome the obstacles which stand in the way of our getting it, and after the obstacles have been vanquished we return to simple satisfaction in the good we have obtained. Thus, passion begins and ends in the concupiscible appetite. In a world where there were no difficulties, the irascible appetite would have no object. In another sense, however, the irascible appetite is superior to the concupiscible. It enables animals, including man, to rise to the challenge of attacking and overcoming difficulties and thus to escape from

being completely determined by pleasure and pain. Man, of
course, can see with his intelligence that it is reasonable to
suffer on occasion for the sake of attaining a greater good. But
it is doubtful if we would actually do this in opposition to our
inclination toward what is pleasant unless we had a sense appetite which enables us to take pleasure in overcoming obstacles.

The fact that the passions of the concupiscible appetite are
hard to order aright needs little proof. Greed for material goods
or desire for sexual pleasure can blind men to what is ordinarily obvious and cause them to act against deeply cherished
ideals. We know too that men want to avoid suffering and
death and that in order to do so they often unjustly make others
suffer and die instead. All this is bad enough, but the worst
moral evils arise when the irascible appetite gets involved.
When it does, men will expend the utmost effort and intelligence in order to overcome any obstacle which stands between
them and some thoroughly despicable goal. In fact, even when
the goal is good, greater evil is often created by immoderate
efforts to obtain it.

The devastating wars of this century, and the threat of even
worse ones to come, have focused the attention of our contemporaries on what is called "the aggressive instinct" of man.
Instinct is a concept which arose originally in connection with
lower animals. Such animals exhibit complex, automatic, un-
learned responses to certain stimuli. This type of behavior is
called "instinctual." It depends upon the existence of physiological structures which are genetically determined and are
not perfected by learning. Is it reasonable then to say that man
has instincts? Montagu claims that man is virtually without
instincts and that in fact one of the important developments in
human evolution has been the replacement of instincts by
reliance on learned behavior.[8]

However, a considerable number of biologists and psychologists have preferred to redefine the term instinct. We can speak
of "open instincts," which are appetitive structures genetically
determined in part but open to modification and development by
learning and experience.[9] This meaning of instinct makes it
approximate the "sense appetite" of the Scholastics. In this sense

of the term man clearly has a "sex instinct." Does he also have an "aggressive instinct"?

Freud ultimately became convinced that human behavior could not be understood solely in terms of *libido*, and so he postulated a "death instinct" which when deflected from its own proper object, the self, and turned outward becomes aggression against one's fellows. Many psychologists have judged that an instinct thus opposed to one's own survival is not plausible, and so there have been attempts to account for aggression against others and against the self as purely developmental and neurotic manifestations. But the prevalence of aggressive behavior among human beings who are not obviously neurotic seems to make this explanation questionable. Today an increasing number of psychologists have concluded that man has, in addition to the pleasure-seeking instincts, not a death instinct but an open instinct for aggression against the environment which, when blocked, can become self-destructive.

Further evidence in favor of an aggressive instinct has come from ethology, the science of animal behavior. Aggressiveness and fighting among conspecifics play a significant role in animal life, a fact which is, at least, suggestive in a consideration of the aggressive behavior of man. Konrad Lorenz and others have drawn the conclusion that human aggression is indeed strongly determined by an instinct inherited from man's animal ancestors,[10] but this viewpoint has been opposed by a number of biological and social scientists who prefer to attribute aggression to social causes.[11] These scientists may well be correct in questioning the validity of the arguments of Lorenz and others as conclusive scientific proofs of their position, since evidence from the field of ethology alone does not appear sufficient as conclusive scientific proofs of their position, since evidence is scantiest precisely in the case of man's nearest neighbors, the primates. Perhaps Lorenz is being misunderstood, but his account of human aggression appears somewhat simplistic and reductionistic; he seems to be neglecting the specifically human aspects of the situation.

But the same charge applies to some of his critics as well.

Social and historical factors are certainly important in the etiology of aggressive behavior, but societies are dependent upon a biological base. How is it that social relationships develop in such a way as to encourage aggressive behavior? The fact that aggression breaks out so readily in the great majority of human groups seems to imply that the biological basis of human societies inclines them to assume forms which make aggression likely.

Philosophical and psychological considerations shed considerable light on this question. Philosophical reflection on the data of common human experience enabled the Scholastics to conclude that man has an irascible appetite, an idea confirmed and elucidated by the data of modern psychology. Within this context, then, the evidence from ethology serves to give further insight. Though culture molds and influences man profoundly, human cultures do not arise in a vacuum; they spring from a biological background and remain rooted in biological structures.

There seems little doubt that man has an irascible appetite or aggressive instinct in some broad sense of the term. What more can be said? Just how dangerous or destructive is this instinct? Perls, Hefferline, and Goodman insist on saying that man has a tendency to destroy. They do this, however, to emphasize the aggressive nature of the drive and do not imply that the "destruction" wrought need do harm to other human beings. On the contrary, the destruction can well be the destruction of obstacles to human understanding or the restructuring of the environment, physical or social, in a way which is of benefit to others as well as to oneself.[12] The irascible appetite or aggressive instinct is primarily a power of self-assertion in the face of challenge and difficulty; it need not be destructive.

Unfortunately, as Lorenz has pointed out, man, like the rat and unlike such "ferocious" animals as the lion, the tiger, and the wolf, has no innate inhibition against killing his fellows and, if neurotic developments are added, can even become sadistic. Furthermore, in intraspecific competition an extra measure of aggressiveness can sometimes be of advantage to an individual or group, even though it is harmful to the species as a whole. Man's lack of inhibition against killing conspecifics results from the fact that, unlike the large predators, man's ancestors did not possess deadly natural weapons. Consequently the species was in no danger of self-extermination. When artificial weapons developed, cultural and moral inhibitions also developed which, until now, have proved strong enough to enable the species to survive and in fact to thrive biologically. This is no guarantee for the future, however. The present non-teleological theory of evolution makes no promises. Rapid cultural advances such as the discovery of nuclear energy create unprecedented situations, and it is conceivable that men may not adapt their behavior with enough speed to survive. A Christian who relies on divine providence rather than on evolution can have more hope for the human race. This does not mean, however, that man's instincts or appetites will not cause him an endless amount of trouble in the future just as they have in the past.

The sex instinct is undoubtedly aroused by external stimuli. Nevertheless there is such a thing as "sexual tension," an internal state which arises even in the absence of external sexual stimuli and inclines an animal to seek an outlet for it. If no suitable object is found, there is a tendency to discharge the tension in ways which would ordinarily be considered unsuitable. Does something similar happen in the case of the aggressive instinct? The passion of anger, one of the more characteristic acts of the irascible appetite, is characterized by a pattern of physiological changes similar in many respects to those associated with sex. In fact, there is a tendency for sexual and aggressive arousal to be associated or to change into one another under some circumstances. This similarity suggests that perhaps the aggressive instinct also involves an accumulation of tension which will tend to discharge itself, even if no suitable occasion occurs. Storr believes that "at the introspective level, it may be true to say that one deplores getting angry; but the physiological changes which accompany anger give rise to a subjective sense of well-being and of invigorating purpose which in itself is rewarding. Appalling barbarities have been justified in the name of 'righteous wrath'; but there can be no doubt that men enjoy the enlivening effect of being angry when they can justify it, and that they seek out opponents whom they can attack in much the same way that cichlid fish do." [13]

It is not clear that Storr is correct in believing that even in the absence of stimulation man would be inclined to aggressive behavior. But where on earth is there an environment which does not induce some measure of frustration in those who inhabit it? In the end one has to say that man's genetic makeup and the nature of the ecosystem in which he lives make aggression likely. Even if a new Adam and Eve were to start afresh on some distant earth-like planet, it seems virtually certain that their descendants would face essentially the same problems as we do. In other words, the present order of the cosmos is such that, without a special gift of God, disorder (concupiscence) will inevitably afflict man in both his concupiscible and irascible sense-appetites.

The fact that man has sense-appetites does not constitute concupiscence. It is the relation of these powers to intelligence and rational desire which constitutes concupiscence. The tragic and painful fact is that the sense appetites are not adequately responsive to reason. Their movement can anticipate that of reason and influence it so powerfully as to blind it. Even if they do not succeed in doing that initially, the passions can pull so hard against rational judgment as to make it seem preferable to yield to them and to act against one's ideals and better judgment. This is what weighs most heavily on the noblest human spirits and constitutes for them an evil beyond any possible physical pain. [14]

This is not, however, an evil which is surprising and unconnected with the basic structure of the universe. It is simply the appearance on the highest level of something which occurs throughout the hierarchy of structures of the cosmos. The higher-order entity is never able to dominate and order totally the subordinate entities of which it is composed. The finality by which the subordinate structures are ordered to the good of the superordinate ones is weak compared to the finality by which they are ordered to their own good. Man's problems are not unique to him. They are the problems of the whole cosmos.

The knowledge which modern science has given us about man should make us aware of how deeply rooted he is in nature. This is true not only in terms of physics, chemistry, and physiology but also in terms of psychology and behavior. Not only does the physical structure and functioning of man grow out of and depend upon the cosmos but also his thought, desires, and moral attitudes. We have seen that there is an aspect of man which transcends the universe; still it remains true that he depends upon and is formed by it. This universe is substantially good, but it is infected with evil. Consequently, not only suffering and death but temptation and moral weakness are inevitable unless God chooses to work a series of remarkable miracles to avoid the consequences of the natural laws He Himself is supposed to have established.

6. SCRIPTURE AND THE PROBLEM OF EVIL

The seeming incompatibility of our present understanding of matter with older interpretations of Original Sin is not of course an apodictic argument any more than the fact that present-day science does not seem to require teleology is an apodictic argument against teleology. If compelling evidence can be obtained from revelation, we shall conclude that the prima facie impression derived from science is wrong and that in all probability the scientific view of the world is due to change; but it does not appear that any such compelling evidence exists.

The two main scriptural sources for the doctrine of Original Sin are Genesis 2 and 3 and Romans 5. As far as Original Sin *originatum* is concerned, Paul has a great deal to add to Genesis. He is undoubtedly one of our main sources for an understanding of the present situation of sinful man, of his need for Redemption, and of the glorious fulfillment of that need in Christ. But today many, if not most, professional exegetes believe he adds nothing to our knowledge of the historical reality of Original Sin *originans* beyond what is given in Genesis. "The general principle may be stated that whether or not a particular ancient text is to be interpreted historically cannot be decided from the use made of it in later texts alone, whatever may appear to be implied, even quite categorically, in the said later text. In fact, the way in which ancient texts are used in the Bible to prove some particular point forms in itself quite a distinct literary genre, of a highly diverse and at times disconcerting kind, so that there is not a great deal that can be safely concluded from such cases until the text which is quoted has itself been examined and considered in the light of its original context." [15]

It has long been accepted that the Bible in general, and Genesis in particular, is not a textbook of science and that the Biblical writers accepted and used the commonly held notions of their own day about the physical universe without intending to teach such views as true in the same sense or with the same authority as their religious message. Hence one can acquire knowledge about scientific facts from the Bible only if those facts are necessarily implicated in the religious doctrine being taught. It does not seem that this is ever in fact the case.

Merely profane history has no better claim than science to support from the *religious* authority of the Bible. But the central concern of the Biblical writers is to reveal to men their personal relationship to God and since personal relationships are essentially historical and include the past as an inner constitutive moment, the Bible is a book of sacred history. Nevertheless this type of history is not very dependent for its essential truth on accurate and circumstantial accounts of past events. The inner meaning of history can quite conceivably be conveyed by means of recitals whose details are only tenuously related to what actually occurred. *A priori* this seems especially likely to be the case in regard to events like Creation and the Fall, which certainly are beyond the reach of any ordinary kind of historical evidence.

It does not seem probable that the author of Genesis had any revealed knowledge of the details of events which took place so long before he or his culture existed. Exegesis, like any human science, must make use of Ockham's razor, and the simplest assumption is that "Israel came to its knowledge of these early events as the result, on the one hand, of inspired reflection on the mighty historical experiences of Jahweh's activity and power through which it had passed, and, on the other hand, of centuries of practical and speculative wrestling with the great problems of life, especially the problem of evil." [16]

God did not correct Israel's understanding of profane history or the material universe. Rather "the often curious ideas which Israel had built up concerning the earliest times—and often enough we can even trace the sources of the material of which these ideas were composed—became the vehicle, the canvas, and lastly also even the indispensable concrete form which God used to communicate the substance of the saving events which it was His purpose to convey. Israel's inspired ideas about the great religious problems and its repertoire of profane information from many different sources have become welded together in a living unity, to form a single narrative. God has grafted His revelation on to Israel's already existing human knowledge." [17]

We must conclude then that the doctrinal content of Genesis is not to be determined by first investigating the historical truth of the various concrete details of the story. The doctrinal content of Genesis 2 and 3 is of much greater importance than its historical content and "the latter can . . . only be effectively demonstrated and established in safety in the measure that its relationship with the former has been shown." [18]

What then were the doctrinal concerns of the author of Genesis 2 and 3? "These two chapters are completely dominated by the *problem of evil*: How is the existence of such a miserable world to be reconciled with God's love and righteousness? How could He have been responsible for creating such a state of affairs?" [19] One part of the sacred writer's answer is clear: God is not responsible for evil. He is both good and omnipotent. The world as it came from His hand was good and without flaw, and it could have remained so. There was no defect or deficiency in it which made the appearance of evil inevitable. Rather evil came by the free choice of creatures.

It must be emphasized that this statement is doctrinal. It does not depend upon historical knowledge but upon insight into the nature and the character of God and man as they had been revealed to Israel. Someone has said that if God's love had not been revealed to us we might perhaps know by philosophical reasoning that God is good but this would not be much comfort to us. For being what we are, the goodness of God might perhaps be more painful and menacing to us than the worst hatred. It was necessary that God should reveal to us that His personal attitude toward us is one of tenderness and loving kindness and that His goodness is also goodness for us. That revelation took place and Israel came to understand that God abhors human suffering, except insofar as it is a remedy for sin. "God did not make death, and he does not delight in the death of the living . . . but through the devil's envy death entered the world . . ." (Wis 1:13, 2:24).

The objection that suffering and death are inevitable by-products of evolutionary progress is certainly a very weak one in the light of the clear teaching of Genesis and the whole of Scripture that God is the omnipotent creator who calls things that are not as though they existed. It is true enough that in the present order of the universe evolutionary progress does not take place without suffering and death. But this is precisely the curse which sin has laid upon the world. The inner dynamism of things has been corrupted and so the evolutionary process which flows from it is also corrupt. Why else does the earth bear thorns and thistles if not because inanimate nature is no longer totally ordered by its finality toward the good of man? This objection has weight only if one insists on conceiving of Original Sin as something which took place on planet Earth after the fundamental dynamism of the world had already assumed its present form. But the clear teaching of Genesis that suffering and death are the result of sin demands that we consider the present dynamism of the world as having been established after Original Sin took place.

How precisely did this come about? Genesis 2 and 3 seem to tell us that man is responsible, but it would be a mistake to regard this as a purely doctrinal statement. Some distinctions must be made. It seems clear that one object of the author was to teach his readers that sin is what makes them unhappy, that by their own sins they are ratifying and continuing the act which brought evil into the world. In other words he is teaching us something about Original Sin *originatum*, about the sinful situation of solidarity in evil which exists now. But what does he intend to teach us about Original Sin *originans*, about the first entry of sin into the world? Sin came because of the free decision of creatures, and the essence of that decision was pride, the desire to be a god unto oneself. This much can be known from the goodness and power of God and from the nature of sin as we experience it. But how could the author of Genesis know that evil first entered the universe on this planet or that it was a human being who was responsible for it? It is true that sooner or later all men implicate themselves in evil and that we experience it as almost a part of our nature. Man is born for trouble, as sparks fly upward. But does this imply that it was the sin of the first man which corrupted our nature or even that there was a first man in the sense of a unique ancestor? It seems clear that it does not. Such notions are part of the popular history which the writer assumed as the vehicle of his thought.

The author of Genesis likes to teach in a very concrete way. He tells us about a particular man and woman and gives us all sorts of detailed information about them. But he also has "a special interest in the origins of all kinds of cultural realities, and for institutions and customs of all kinds, whether religious or secular. In his treatment of these he follows a striking method, which is to relate how and why such and such began as if it were an actual event which occurred at a particular moment in the past. He does this in the familiar anecdotal style to which we are accustomed in folklore. These episodes are of course most often more a kind of dramatic characterization of the thing in question, as it is known by experience, than actual history of the past. In the same way, the numerous explanations of names are also a familiar teaching procedure of the same kind." [20]

What we would convey by the abstract statement that human nature has been corrupted from the ideal pattern which God intended, the author of Genesis conveys by a story about the corruption of a particular concrete man who carries all other men in his loins. Thus it is no part of the doctrine of Genesis that the Fall of the universe in general and of man in particular from the state intended by God took place on this planet. The whole apparatus of the garden was derived by abstracting away the evil from the world we experience now. It never existed or could exist. Nor could the universal man "Adam" in whom we were all contained and whose fall meant

168   COSMOS

our own corruption and death ever have existed here on earth. It seems rather that he is a dramatic concrete representation of a potentiality hidden in the ground of the universe, a potentiality which when corrupted achieves a flawed and imperfect actuality not only in our own race but in all races of intelligent beings in the universe. By means of the image of a first unique Man, the author of Genesis 2 and 3 has telescoped into one dramatic scene the Fall which long ago corrupted the entire universe and its later impact upon our own race.

7. THE FALL OF THE COSMOS

What then are we to say about this mysterious event, "the Fall," which the eyes of the inspired writer dimly discerned through the mists of nearly countless aeons? For us the Fall must be not merely the Fall of our own biological race but rather the Fall of the entire universe. The event which twisted the world from the path God had set for it must have happened long before Earth existed, in all probability before the evolution of the elementary particles some ten billion years ago.

If the Fall occurred long before our race existed, who was responsible for it? Certainly the being or beings in question were created long, long ago and held a position in the universe far more central than ours. Perhaps they were nevertheless incarnate spirits, men. But the simplest supposition is that they were not men at all but rather the persons whom the book of Wisdom tells us are responsible for death, the Adversary and those associated with him.

Paul explains the universal salvific role of Christ in terms of opposition to Adam. But this is not the most common or most prominent way in which the New Testament presents it. Both the Synoptics and the Johannine literature oppose Christ to Satan. In the Synoptics the mission of Christ begins with His struggle with Satan in the desert, and throughout His ministry He shows Himself as the one who is bringing Satan's dominion over the world to an end by establishing the kingdom of God. According to St. John, the head of sinful mankind is not Adam but Satan (Jn 8:44), and the passion is the hour of the "prince of this world" in which he tries, through Judas and the "Jews," to destroy Christ, but instead is judged and cast out (Jn 12:31).

The Evangelists and the New Testament as a whole present Christ's work as a struggle with superhuman "Principalities and Powers" who are the rulers of the present age and the source of its evils.[21] This way of viewing the life and work of Christ is far more fundamental than the Christ–Adam polarization and probably goes back to the Lord Himself.

Even if one feels that it is necessary to retain the idea that the "Fall" of man is an historical event which took place on this planet because of the sin of a member of our biological race, one would still have to admit that this event is a secondary one and that there must have been a previous "Fall" in which the Principalities and Powers rebelled against God. This first Fall is the main source of the evil of the present order of the universe and, in fact, suffices to explain it. There is no need to invoke a human ancestor besides.

In this case why did St. Paul explain the role of Christ in terms of opposition to Adam, a human being, rather than in terms of opposition to the Principalities and Powers and their leader, Satan, as did John and the Synoptics? I believe that he was more concerned than they to emphasize the physical solidarity in sin of man without Christ. For the Biblical writers generally, but especially for Paul, "flesh" is the totality of human nature insofar as it is subject to weakness and mortality and is separated from the holiness of God. It is also a principle of solidarity. In the Bible generally, "flesh designates kindred in a very concrete sense; all the members of a single kinship group have one flesh, which is conceived as a collective reality possessed by all." [22]

Paul thought of Christ as entering into this solidarity of "sinful flesh" (Rom 8:3) by being "born of a woman" (Gal 4:4). Thus God the Father "made Him to be sin who knew no sin" (2 Cor 5:21) in order that He might "condemn sin in the flesh" (Rom 8:3). Christ became "a curse for us" (Gal 3:13) in order that we might be reconciled to God "in His body of flesh by His death" (Col 1:22). The condemnation or destruction of sin and the reconciliation to God are, of course, complementary aspects of one process brought about by the death and resurrection of Christ "who was put to death for our trespasses and raised for our justification" (Rom 4:25). With a hammer blow God the Father has shattered the old order of sin and death as it existed in the flesh of His son and transformed Him into "lifegiving spirit" (1 Cor 15:45). Thus in Christ the universe which was "subjected to futility" has been "set free from its bondage to decay" and has attained "the glorious liberty of the children of God" (Rom 8:18–25). But its complete transformation still waits for our own death and resurrection and the transformation which we must yet undergo. For "flesh and blood cannot inherit the kingdom of God, nor does the perishable inherit the imperishable" (1 Cor 15:50).

Thus, in spite of his use at times of forensic categories and his personification of sin and death, Paul's conception of the Redemption is very "physical." This is more often pointed out, perhaps, in regard to the life of the justified man in the body of the risen Christ,[23] but it is just as true of the state of fallen man. In order to emphasize the notion of a real solidarity in sin and death, Paul, following the author of Genesis 2 and 3, made use of the most powerful and realistic image he or his culture knew, the biological unity of the "flesh." Today we need to retain his profound insight into the physical solidarity in sin and death of the present world order, but we need to express it differently.

Here I think some writers on Original Sin who well realize the necessity of taking contemporary knowledge of the world into account fall short. Either, like Teilhard, they come perilously close to denying that the present world order is a solidarity in sin and death which could not have been created by God and needs a Fall to explain it, or else, like Schoonenberg,[24] they make the solidarity too feeble. Man is not enmeshed in a situation of sin simply because of free decisions culturally transmitted. Such a predicament could be overcome by simply reversing the decisions.

Rather, sin and death are written into his very genes in such wise that a sinless human life was from the very beginning a miracle of heroism which could not be expected from a large group. In fact, sin and death are even more deeply written into the universe than that. Not only the structure and functioning of DNA molecules implies sin and death but even the structure and functioning of atoms, elementary particles, and whatever structures underlie these. The author of Genesis tells us that man's world is "cursed." To previous centuries that meant the planet Earth and whatever fundamental processes they knew, e.g., the reproductive process. For us that means the entire universe and the behavior of matter down to its most elementary structures, which are the roots from which the macroscopic events we experience bud forth. Our present world is the result of an evolutionary process which is probably ten billion years long and which went awry from the very beginning. As a result, physical entities on every level are partially isolated from one another and pursue their own individual good regardless of the effect this has on others. Hence they often destroy one another and the higher wholes of which they form a part. In consequence evolution is not simply a harmonious unfolding of an intrinsic finality which aims at the good of the whole universe but a slow movement amid a maelstrom of conflicting forces which arrives finally at precarious equilibrium at the price of many false starts and much wastage. On the animal level we see the conflict, pain, and death which made Blake wonder who had formed the tiger's fearful symmetry. On the human level this tortuous process finally becomes murderous, a perfect expression of the interior state of rational creatures isolated from God's love and from one another.

Paul, like John and the Synoptics, knew that this sinful world-order is both the result and the symbol of personal decision. He gives more prominence than they to the role of the symbolic figure Adam, but even for him it is "the rulers of this age" who "have crucified the Lord of glory" (1 Cor 2:8) and it is against them, not "flesh and blood," that we ourselves are now contending (Eph 6:12). Again, it is Paul who tells us that the whole of creation "was subjected to futility" and "has been groaning in travail together until now" (Rom 8:20, 22), an idea which suggests that he saw considerably more evil in the world than could be caused by the sin of a man, even the first man.

Within the perspective outlined above, the reflections of

recent writers on human sinfulness find a natural place. For example Schoonenberg's excellent analysis of "the sin of the world" seems far more acceptable if it is understood that the whole drama is played out on a plane which was already determined by the primeval fall of the whole universe. Man was already doomed to a history of sin when he first came upon earth. The question then could only be, how much sin? If he had not gotten so entangled in it, he might have been able to accept Christ and perhaps, as Guardini suggests,[25] escape it altogether. That he did not was a tragedy; however, it has not left him in a worse condition. It merely means that his way of escape must now be the only route he left open to Christ, namely, the cross.

8. THE REBELLION OF MODERN MAN

Theories of Original Sin which were prevalent before Darwin were inadequate, but they embodied revealed truth which man needs in order to make sense of the world. If that truth is rejected, there is no way for man to understand his existence, and the strain on his trust in God's goodness is increased to the point where many are likely to reject faith entirely. Unfortunately, the effect of modern scientific knowledge about the structure of the cosmos has led many to suppose that the concept of Original Sin is simply a mistaken notion which should be eliminated.

This attitude is not sufficiently respectful of tradition. We certainly cannot afford simply to accept tradition as it is handed to us and to repeat it word for word in an uncomprehending manner. But neither can we afford to reject it without long and serious consideration. Our first supposition ought to be that tradition expresses, albeit imperfectly, a truth which we need to know, and if there are inconsistencies in it, either internal or in relation to well-established scientific truth, then we ought to suppose that it can be reformulated in a way which will make the revealed truth embodied in it clearer and more intelligible. In the case of the doctrine of Original Sin the penalty for not reformulating tradition seems to be the error of absolutizing the present world order, and this obscures both the omnipotence of God and the full reality of created freedom. It is not any lack of power on the part of God which allows evil to come into existence; it is rather the fact that His power is so great that He can create beings who themselves are creative. The present world order need not have arisen; it was creatures, including ourselves, who would have it so. Modern man, who shares with Satan the guilt for the present situation, has tried to exculpate himself by throwing the blame on God, the one to whom sin is most offensive.

The classic argument in favor of atheism—really the only argument—was stated by Epicurus and quoted by Hume: "Is he willing to prevent evil, but not able? then is he impotent. Is he able, but not willing? then is he malevolent. Is he both able and willing? whence then is evil?" [26] In modern times the same thought has led to what Camus calls "metaphysical rebellion" on a wide scale. "Metaphysical rebellion is a claim, motivated by the concept of a complete unity, against the suffering of life and death and a protest against the human condition both for its incompleteness, thanks to death, and its wastefulness, thanks to evil. If a mass death sentence defines the human condition, then rebellion, in one sense, is its contemporary. At the same time that he rejects his mortality, the rebel refuses to recognize the power that compels him to live in this condition. The metaphysical rebel is therefore not definitely an atheist, as one might think him, but he is inevitably a blasphemer. Quite simply, he blasphemes primarily in the name of order, denouncing God as the father of death and as the supreme outrage." [27]

What are the motives which have led our age to rebel? No doubt they are many, ranging from simple misunderstanding to Satanic pride. The occasion for them is the fact that no solution to the problem of evil is fully adequate. The wheat and the tares grow together till the end of the world. We know that God hates evil and did not cause it; we know that He permits the normal consequences of evil acts because from them He is drawing a greater good for those who love Him. But though we can speculate about what that good may be, we do not see it clearly. Ultimately we must trust Him and be willing to suffer with Christ in order that we may also be glorified with Him (Rom 8:17).

But Christians cannot escape part of the responsibility for modern infidelity. "Love alone can be believed" [28] and our failure to display more clearly both in our lives and in our theories that all God's actions are manifestations of love has in part occasioned it. The practical failures have no doubt done the most damage, but inadequate theorizing, even by outstanding Christians, has also played a role. Let us imagine Teilhard de Chardin accompanying Dr. Rieux to the bedside of the dying child in Camus' novel The Plague. Dr. Rieux is a sensitive man who is capable of a good deal of real love. Teilhard, however, was something more—a man of deep holiness—and so I see him being even more deeply affected than Rieux, even more willing to sacrifice himself in order to alleviate the sufferings of the child. Later, however, as they walk away from the house Rieux breaks out into his famous tirade against the notion of a God who would permit such evil and Teilhard airily replies: "My dear Rieux, don't agitate yourself so. The sufferings you have witnessed are nothing more than 'so many by-products (often precious, moreover, and reutilisable) begotten by the noosphere on its way' to the Omega point." [29]

Would Teilhard actually have spoken so? Of course not. Aside from the fact that he was a saintly and tenderhearted man, he was well aware of the true nature of evil, having experienced more than his share of it. His casual dismissal of evil took place on the theoretical plane only. Thinking in a more concrete mode he wrote: "At every moment the vast and horrible Thing breaks in upon us through the crevices and invades our precarious dwelling-place, that Thing we try so hard to forget but which is always there, separated from us only by thin dividing walls: fire, pestilence, earthquake, storm, the unleashing of dark moral forces, all these sweep away ruthlessly, in an instant, what we had laboured with mind and heart to build up and make beautiful." Against "the temptation to curse the universe, and the Maker of the universe," his only defense was to "adore it by seeing you [God] hidden

within it" and to believe that "the immense and sombre Thing,
the spectre, the tempest—is you [God]." [30]

Here we have the crucial decision. To overcome the temptation to infidelity Teilhard, like innumerable other Christians,
disregards his own intuition of the nature of evil, that "vast
and horrible Thing," and ends by identifying it as a sacrament
of God. To which Rieux (Camus) and innumerable other
moderns reply, that sacrament shows me the true nature of
your God, and so I despise and reject him.

Teilhard has not evaluated properly the kind of concrete
human understanding which is often referred to as "feeling."
What we call "feeling" is an act of symbolizing activity in
which a complex pattern of activity proceeds from the personal
subject who has been actualized by contact with the real.
Often enough our training in how to feel is so defective that
we do it badly, and this kind of human activity is even more
subject to the distorting influence of sin than is abstract in-
tellection. Still it is a central kind of human act and gives us
real knowledge of the world which is inaccessible to abstract
thought. Teilhard and others would have produced better
theories if they had paid more attention to the deliverances of
their "feelings" about evil.[31]

Camus remarks that "in the Western World the history of
rebellion is inseparable from the history of Christianity," because "the only thing that gives meaning to human protest is
the idea of a personal god who has created, and is therefore
responsible for, everything." [32] God is, of course, responsible
for everything, but not in an unqualified sense. It is the failure
to note the qualifications which makes Teilhard's viewpoint
on evil (which, as far as the essential point goes, is that of Father Paneloux in The Plague) so potentially dangerous.

Camus himself operates with the same false presupposition
as Teilhard. As Magee observes, he falsely identifies the will
of God with the present order of nature. The world is not the
measure of God; rather God is the measure in terms of which
the world is found wanting. "Camus's mistake, then, is to
turn from faith in a 'scheme of things' to faith in man, in
sympathy for fellow sufferers, and in the courage that fights

plagues without hope of ever finally overcoming them. . . .
But this is merely to return to one of those discredited features
of the world—the presumption of human constancy. If the
light of existence has really gone out—if evil has done its terrible and salutary work of disillusionment—how can one turn
sentimentally to the fragmentary and inconstant human love
that we know is so infected by selfishness and fear?" [33]

It may be that Camus himself and other contemporary
agnostics of good will are not turning to man merely as a part
of the order of nature but rather to man as a symbol of the
transcendent God to whom he is directly related. Indeed, if
their putative good will is really good, that is necessarily the
case in spite of the defective way in which they explicitate,
even to themselves, their primordial knowledge and desire. But
hope in man as such is doomed to disappointment.

9. THE EXISTENCE OF COSMIC POWERS

The discussion of sections four, five, and eight has displayed
further reasons supporting the position taken in section three,
namely, that evil is deeply rooted in the cosmos and affects
even the material structures which lie below the human level
and antedate the appearance of man on this planet. The acceptance of this proposition leads, in accord with the argument
of section three, to the conclusion that cosmic process is grounded
in the intelligent activity of created being(s) who are, along
with man, the cause of evil. This conclusion, which does not
depend essentially upon strictly theological considerations
(though in this book I have integrated it with them) buttresses
the argument of section seven, which is weakest precisely in
regard to the existence of the superhuman spiritual beings
known traditionally as angels and demons. It can be argued
that the belief of the Biblical writers in such beings was merely
cultural, a supposition which they adopted from their environment and which does not form a part of the revelation which
they announce to us. This hypothesis does not appear terribly
plausible to me. As I have noted, whatever may be said about
the role of angels in the Old Testament, the existence of the
Adversary is presupposed by one of the strongest New Testa-

ment themes, the conflict between Christ and Satan. To change this into a struggle against impersonal evil and collective human sin requires radical surgery which does not commend itself to me without more evidence than seems to be forthcoming. Nevertheless the confluence of the philosophical argument of section two with that of section seven establishes the conclusion in greater security. This is welcome because this conclusion is something of a scandal to the modern mind in view of its deeply rooted heuristic anticipation that the cosmos must prove to be an impersonal world machine.

The notion that personal spiritual principles lie behind the visible material structures of the universe was congenial to the minds of the Biblical writers and of the ancient and medieval philosophers. It has also been congenial to some outstanding contemporaries, such as C. S. Lewis and Karl Rahner. In his essay on the theology of death Rahner develops some notions about the role of the angels in creation and about the meaning of Christ's "descent into Hell" as the reversal and destruction of the Devil's influence. Following St. Thomas, he hypothesizes that the angels play a role in the constitution of the cosmos. They are principles of the world, "the ultimate foundations of the natural order of things, determining the right order of events in this world because of their essential relationship to the universe." [34] But such views have lost favor in modern times because of the rise of science which has proved capable of explaining a great deal without recourse to that hypothesis and, Christian readers should note, without recourse to the hypothesis of God as well. But today we are coming to realize what should have been obvious all along, namely, that science does not deal in ultimates, not even ultimates of the natural order. No matter how accurate and general a description of material structures it may give, the problem of their ground will still remain. Science describes the structures which emanate from that ground but makes no attempt to read in them what they symbolize. That is left to men (including scientists, of course) who reflect on their ultimate personal concerns.

Christians will see little reason for rejecting this bias of the modern mind in favor of an impersonal cosmos unless they see the connection it has with the problem of evil. In section

eight of this chapter I have tried to stress as strongly as possible
the correctness of one of the essential insights of Camus and
other modern agnostics. The present world order is inextricably
entangled in real evil, and if God is its sole creative cause, then
He must be held responsible. Ironically, of course, the very
standard by which the nonbeliever arraigns God is the imprint
within him of the divine goodness. The rejection of supposed
divine injustice is an implicit affirmation of God's love and pity
for a world which has gone awry.

10. A HAPPY FAULT?

The evil which the Adversary and other sinful creatures, both
angels and men, have done cannot be annihilated. It exists and
must stand forever. However, God could, if He wished, counter-
act it by intervening miraculously in the workings of creation.
The traditional doctrine of the preternatural gifts asserted that
He was in fact willing to do so, at least as far as man was concerned, but this does not seem to be correct. The parable of
the wheat and the tares (Mt 13:24–30) may perhaps be given
a cosmic interpretation parallel to the ecclesiastical one of
Matthew 13:36–43. The good which God wants to create
requires that evil be allowed to take its course. Angelic and
human freedom can be actualized fully only if the alternatives
it faces are real. Our lives are serious and can be great because
God has given irrevocably to creatures the power to be real
causes of great good or great evil. Miracles do happen, but the
amount of physical good they do is negligible in the face of
the vast suffering of creation. They are the eschatological signs
of the divine love and pity and of what will occur at the end of
the present age when creaturely freedom shall have worked
itself out fully.

Though He is unwilling to contravene the order of creation
miraculously, God has nevertheless intervened in an unexpected
way. Satan, "a murderer from the beginning," was not mistaken
in thinking that he could mortally wound the original hopes of
God for a sinless and happy creation, but he was not aware
that through the incarnation and death of the Son of God

death could be made the gateway to life. From vast evil an even greater good has arisen, the glory with which the Father has endowed Christ and those who will follow Him.

By this, I do not mean that the cosmos as a whole is better than it would have been if sin had not occurred. As it came from the hand of God creation was wholly good and without flaw. Evil entered it because of sin, and God does not want, is incapable of wanting, sin. He desires that His creatures should love Him and seek Him and thus attain their perfection. He does not desire this for His own fulfillment but for theirs. Sin is what opposes and thwarts, insofar as this is possible, the divine will from which all good flows. It is therefore self-contradictory to say that creation can be improved by sin. Those who speak in such a way are confusing God with man. In human affairs unexpected problems and failures can lead to better performance and greater satisfaction. But this happens because man's foresight and power are limited and because challenges can inspire us to greater effort. The divine work of creation would have been best if it had developed exactly as God wished. For what He wills is good, and good is what He wills.[35]

Nevertheless, there is a millennial-long custom among Christians of calling Adam's sin a "happy fault." What could be the meaning of such a phrase? It may very well be partly based, as the Holy Saturday liturgy suggests, on the erroneous notion that Christ would not have come had sin not occurred. If that were true, the universe would be better off because of sin— but this implies only that in fact the Incarnation was willed independently of the decisions of creatures. Must we conclude that this manner of speaking is simply false? I would prefer to believe that it expresses a truth. If so, it would seem that the truth can only be this: While creation as a whole has indeed been harmed by sin, those who attain salvation through Christ will be better off than they would have been. We might compare man's situation to that of soldiers going into a war which has been forced upon them. If they survive, they will be better off because of their victorious struggle than they would have been were there no war. But obviously this does not mean that the war is a good thing. Neither would anyone want, even if he knew that he was going to survive, to purchase personal development at the price of others' blood.

I think it is probably true that in the present order of things those who are faithful and are saved will be just as well off as, or even better off than, if sin had never entered the world. At least this will be true of the Virgin Mary and the great saints. But it is impossible that the total situation be as good as it would have been without sin. All the extra good which has come to creatures in the present world order is not worth the eternal loss of a single spiritual being. It would be immoral for a man to hope for the damnation of anyone in order to secure any good whatever. But if it is immoral to do so, that is because it is against God's will. If it is against God's will, that is because He judges it evil to do so, and if He judges it evil, it is evil. In this sorites I switch back and forth between considerations of intelligibility and being. But that is legitimate because ultimately intelligibility and being are identical. God Himself is the source, norm, and prime analogate of both. What He hates is evil, and evil is what He hates.

Nevertheless, from the freely chosen evil of creatures God draws benefit for those who are faithful to Him. The evil in the world can indeed make us suffer, but in the end evil will work for our greater good—provided we are faithful. "Now we know that for those who love God all things work together unto good" (Rom 8:28). Christ has made up the difference with His blood.

NOTES

1. BERGER (I), p. 80.
2. Nevertheless there is nothing completely new under the sun (at least in philosophy), and the point of view to be developed would not have appeared so strange to Gnostics, Manichaeans, Neoplatonists, and many Church Fathers. Indeed it is a nice historical problem to explain why the point of view which I espouse has been so little considered in recent centuries. I believe it is a modern form of a perennial philosophical and theological option.
3. FEYNMAN, p. 14.
4. ELIADE, p. 77.
5. PENDERGAST (I), (II).
6. DOBZHANSKY, p. 200.
7. For a brief account of Thomistic doctrine on this point, see GILSON, Chapter 8.

8. MONTAGU, Introduction, especially p. xii.
9. STORR, Chapter 2.
10. LORENZ; ARDREY (I), (II); STORR.
11. MONTAGU.
12. PERLS, especially pp. 340–342.
13. STORR, pp. 18, 19.
14. The lamentations of a St. Augustine or Teresa of Avila over peccadilloes almost invisible to most of us are mainly the result not of neurosis but of an acute appreciation of real values. Just as in the scientific sphere there are people like Einstein who possess a sensitivity to intelligibility which seems supernatural compared to that of the ordinary professor of physics, so in the moral sphere the saints have a sensitivity to values which far surpasses the ordinary.
15. RENCKENS, p. 102.
16. RENCKENS, p. 41.
17. RENCKENS, p. 43.
18. RENCKENS, p. 247.
19. RENCKENS, p. 158.
20. RENCKENS, pp. 256, 257.
21. Cf. SCHLIER, passim.
22. McKENZIE, art. "flesh."
23. FITZMYER, p. 824, no. 140.
24. SCHOONENBERG.
25. GUARDINI, Part III, Chapter 10.
26. MAGEE, p. 427.
27. CAMUS (II), p. 24.
28. BALTHASAR, p. 83.
29. TEILHARD (I), p. 313.
30. TEILHARD (III), p. 90.
31. Let me emphasize that although I am criticizing Teilhard's position on the problem of evil I have great admiration for the man and his work.
32. CAMUS (II), p. 28.
33. MAGEE, p. 442.
34. RAHNER (I), p. 32. See also RAHNER (II), Chapter 10.
35. Note the connection between the position on evil developed here and my earlier denial of predestination. If God in no way predetermines creatures to sin but sets up a structure of possibilities in which sin results in greater good, then we have to envision Him as hoping that His love will be rejected in order that He may do greater good. But the love which is rejected is precisely that from which all good comes. The whole thing seems hopelessly self-contradictory to me. On the other hand, if predestination is assumed, we fall into all the difficulties associated with that concept, among which is the notion that God causes that which is opposed to His will, i.e., which He does not want to happen, to happen.

8

Hope

1. THE FUTURE OF MAN

WHAT OF THE FUTURE? The cosmos continues to develop throughout the enormous extent of space. The universe is perhaps ten billion years old. Galaxies are still being formed, and it may not be unreasonable to suppose that cosmic history will extend as far into the future as it does into the past. Man has existed for about a million years, Christianity for two thousand, and science for four hundred. If we consider the history of the cosmos to date as the first "day" of creation, then man is something like ten seconds old, Christianity two-hundredths of a second, and science three-thousandths of a second. In all probability the history of man has barely begun.

Today man stands poised on the edge of space. Perhaps the term "poised" is a misnomer, for our position is certainly not a stance of secure mastery. We have managed to reach the moon with the expenditure of great—possibly excessive—effort. We, no doubt, will eventually explore other parts of our solar system, and I am convinced that, ultimately, man will reach the stars, either bodily or by exchanging information with other races who already live there. If the world were absurd, there would be no reason for holding that opinion. But the world, though marred by evil, makes overall sense. It is an artifact and symbol of Cosmic Powers who constructed it purposely on such a huge scale. It seems impossible that the vast extent of the universe exists solely for the sake of planet Earth, a mere dust speck amid the galaxies. Neither man nor the artificers of the universe are ultimately interested in expanses of inanimate matter which have no relationship to intelligent life. One of Piaget's children when asked why there were two mountains, one large and one small, in view of his home replied that the little one is for short trips and the big one for long trips. The conceptual expression was naïve, but the basic insight was correct: the universe exists for the sake of intelligent life; it is ours to explore. No one can guess the nature of the scientific developments which will make this exploration possible, but in a billion years the human race will know much more about nature than it does now.

It may well be, of course, that man will not have to discover everything for himself. The evolution of intelligent life has probably occurred or is occurring on many planets besides Earth. MacGowan and Ordway attempt to estimate N, the number of intelligent societies which have evolved within our own galaxy to the point where they are capable of interstellar communication. They summarize their estimates with the following pedagogical equation:

$$N = N_s \cdot f_s \cdot f_p \cdot f_l \cdot f_i \cdot f_c.$$

N_s is the number of stars in our Milky Way galaxy suitable for the development of higher forms of life; f_s is the fraction of these stars having planetary systems; f_p is the fraction of these systems having at least one planet suitable for the development of higher forms of life; f_l is the fraction of these planets where such life has actually developed; f_i is the fraction of these planets where life has achieved a human level; f_c is the fraction of these planets on which interstellar communication has been achieved. After discussing and estimating the factors on the right hand side of the equation, they arrive at $N = 3 \times 10^9$. [1] Some of their assumptions appear naïve and some may be grossly inaccurate, but when one considers that there are about ten billion galaxies observable with present-day telescopes and that undoubtedly many more exist, one must conclude that what we know indicates that the number of intelligent races which exist, have existed, or will exist in the universe is quite large. In regard to reports of flying saucers the only intelligent attitude is practical skepticism; if there is no convincing evidence, we assume that it did not happen. But at the same time we must realize that it may have happened and, if not, that it could happen at any time.

Is it plausible to think that there are many intelligent races in the universe and to believe at the same time that our race is the most technologically advanced among them? It might be pleasant for a Christian to believe that the birth of science can happen only in a society fertilized by the Judaeo-Christian revelation, but one must reflect that there were many ancient peoples more culturally advanced than the Israelites. Consequently, while the uniqueness of Christ makes it possible for a Christian to conceive that his race may be uniquely advanced, it is also conceivable that it is not. We have no way of knowing whether our own efforts at the exploration of the universe are unique and whether man will bring the technological gospel to others or vice versa. But if there are indeed other races in the universe, they will probably make contact with one another sooner or later, either bodily or by exchanging information. The cosmos is one. The Cosmic Powers on the side of God form one community, and the universe is their common project. They share the design of Christ to unite the scattered children of God, that is, all intelligent beings in the universe, all of whom are made in the image and likeness of the one exemplar.

What is the role of our race in this project? Whether our science and technology is unique, the revelation entrusted to us in Christ certainly is. There is only one Son of God who entered the cosmos at a unique time and place. We have no more reason for exalting ourselves on this account than the ancient Israelites had. As their prophets told them, Yahweh did not choose them because they were the greatest of peoples but because they were the smallest. They were to be the bearers of a message greater than they, and their greatness was conditional upon humility. The race of man stands in a similar position. Our

186 COSMOS

own achievements cannot possibly equal the value of the
revelation entrusted to us. It is possible that Christ chose to
become a member of this race because it is one of the more
sinful and needy ones in the universe. In a billion years the
planet Earth may well be forgotten except for one fact: on this
bit of cosmic dust the Son of God entered the world.

2. THE SHAPE OF THE FUTURE

What will the men and the societies of the future be like?
The Australopithecines were probably human persons and,
therefore, our equals in what is essential. Nevertheless their
intellectual abilities and their culture were so far inferior to
ours that it would have been difficult for them to imagine us
even if they had known that an evolution was to take place. If
human history is to continue for another billion years, it may
be that the men of that far distant future will be unimaginably
different from us. Obviously the same may be said of the other
races which may exist in the universe. On the time scale of
the cosmos a few million years is very short. Somewhere there
may be a race a few million years older than our own which
has taken charge of its own evolution and has already progressed
to an unimaginable state of development.

We cannot know the path which development will take in
our own race or in other races, but we can speculate, and our
speculations could be right—at least in some respects. The most
likely path for further development is cultural change, for
such change has been most prominent in human history during the past ten thousand years. There is probably very little
difference in biological endowment between the men of twenty-
five thousand years ago and us, but our advanced culture gives
us enormous advantages over them in certain respects. We can
expect cultural advance to continue, and it is probable that the
greatest changes still lie before us.

As we have seen, the concept of an intelligent machine is
self-contradictory, but the related concept of an intelligence-
amplifier is another matter. The aid given to human thought by
the invention of writing was enormous. Perhaps the discovery
of electronic information-processing will prove even more sig-

nificant. The ability of machines to store, manipulate, and transform large masses of data in complex ways will certainly be increasingly more useful and will enable us to effect intellectual projects which otherwise could not be done. Perhaps the availability of this power will change human thought as much as machines have already changed our powers of locomotion. It is conceivable that some day it will be possible to make direct connection between the human nervous system and machines which would enable men to communicate with and use any kind of machine simply by thinking about it.

Deliberate genetic improvement offers another possibility for the continuation of evolution. At present we do not know enough to undertake such a program, but, as the science of genetics develops perhaps it will be possible to change man's genetic endowment, either by control of reproduction or by the more radical means of direct alteration of the coded information. If this occurs, some of the changes may be directed toward facilitating symbiosis between man and his machines.

It is conceivable that intelligence may be capable of embodying itself in structures quite different from our own bodies. The speculations of science-fiction writers about such material configurations have ranged from computer-like artificial structures to solar magnetic fields to as yet undiscovered topological structures of space itself.

There is a certain affective ambivalence about all such speculation on the future of intelligent life in the universe. On the one hand it opens vistas and can help us to sense the wonder of the unknown possibilities which still lie hidden in the world; on the other hand it can generate a nightmarish psychological aura. One shrinks from the prospect of the ultimate dehumanization by which man is stripped of his very nature. Such horror reflects a valid insight. Heretofore man was normalized by his immersion in a universe whose structure he did not create. His powers of self-manipulation, and therefore his powers for self-stultification and self-destruction, were limited. Now, however, he is about to assume more responsibility. The danger inherent in this can be understood by simply walking through the heart of a large American city with mind and senses open to the full reality of the situation. Man could conceivably create for himself a human world which would be an image of hell and make himself into a true child of Satan.

If man were dependent on man alone, such fears would be more than justified. But though there is little hope in man conceived of as totally autonomous, there is good reason for hope in man the child of God and the brother of Christ. The patterns which matter and history can assume have not been determined mainly by Satan nor can they be utterly changed by sinful men. Intelligent life can only take certain forms, and the horrors imagined by some science-fiction writers are largely chimerical. Furthermore, even within the realm of the real possibilities immanent in the original form of the cosmos, the worst have been eliminated by the entry of Christ into the world. The world we now live in is a redeemed one whose created ground has been transformed by the presence within it of Christ and of the Spirit which He has poured out upon it. Consequently, the history of the cosmos and the future development of intelligent life within it must be substantially the history of divine love working itself out in time. Until the day of judgment there will be no radical discontinuity with the present, and if human life is indeed destined to assume forms unimaginable to us, those forms will nevertheless be ones chosen, in the main, by love. The Australopithecines, while regretting as we do the weight of evil with which history is burdened, nevertheless take satisfaction in the way in which life has developed. If there is to be further development, we too one day will take satisfaction in it.

The future may bring forth remarkable developments, but these developments will remain on essentially the same plane of existence. The worth of a man is not determined mainly by the material structure in which his spirit is embodied but by his personal relationship to God and his personal decisions. The genetic and cultural endowment of contemporary man is quite adequate for a life of the highest nobility, as is evident from the fact that Christ chose to come some 2000 years ago. Indeed, this fact makes me skeptical of the idea that intelligent life can assume a form much different from ours. Even though I anticipate a human culture which will eventually extend over the whole universe and will be much more advanced than ours in some respects, I shall be surprised if the level of individual intelligence displayed by the author of the Gospel according to John is ever exceeded.

3. THE END OF TIME

Christians have traditionally believed that history had a beginning and that it will have an end, after which the cosmos will enter into a state of fulfillment in which change either will be absent or will have a different character. The conception of being and symbolic activity which we have developed sheds light on this. Being is either infinite, in which case it is God, or finite and limited. Finite being needs only a limited period of development in order to express and fulfill its primordial potentialities, and so its history has a beginning and an end, after which it enters into a state of fulfillment which is not temporal in the same sense as its period of development.

In the case of an individual there is no problem about this. Each of us is aware of himself as a definite person who remains the same through all the events and developments of a life history. Our essential self is identical with the primordial subjective knowledge and desire which is the term of our constitutive relationship with God and also with the essential unchanging structures of the cosmos. Man's primordial potential for development is his essential personhood, his "name," which all the effort of his life expresses and actualizes.

The cosmos as a whole is another matter. One can conceive that God might continue creating new persons indefinitely so that, even though each individual would have a finite period of development, the society of all the persons in the cosmos would grow forever. I assume that the traditional belief in a definite end to the present aeon means that this will not happen. The creative intention of God envisages a definite society of persons centered on His incarnate Son, and the definite potential of this society will in time be actualized and expressed, after which will come the end of this age and the beginning of the new creation.

Development in the fullest sense of the term will necessarily be excluded from this definitive state; succession, however, will not. The self-possession of finite subjects is not total; hence they cannot express themselves fully in one act and must make many different acts. During the period of development, the cosmos and each individual subject expresses something new in at least some of their acts. When development is completed, new moments of time (i.e., new restricted presents) will continue to proceed from the past, but they will not contain anything essentially different. Activity will be pure enjoyment of what is already possessed, a kind of endless and exalted play before God. In Piaget's terminology, perfect equilibration will have been achieved, and radical accommodation will no longer be necessary.[2]

Thus there are three different "modes" of being defined by their different relations to succession. The first is eternity. Eternity is the denial of succession. It is proper to the infinite divine persons who possess themselves totally and are able to sum up and express their full reality in one infinite act which does not develop but simply is. The second is temporality as we experience it. This is proper to a developing being who is not only unable to express himself fully in a single act but who also has not yet expressed some aspects of himself. The third, which we can call "sempiternity," was described in the preceding paragraph. It is proper to limited beings who have already expressed themselves fully and thereby have produced structures which embody the totality of their primordial potentiality.[3] Having completed their development, they now use those mature structures to produce mature and perfect acts which collectively and successively re-express the fullness of their personalities. To arrive at this third state in which activity is pure and effortless enjoyment is the object of the striving of beings in the second.

4. THE FINAL WORD OF GOD

How will the cosmos in general, and each man in particular, arrive at its consummation? Three general patterns are conceivable. One is development which marches triumphantly to its goal, a progress not without difficulties but finally triumphant over all obstacles. Such is the progress of a Greek demi-god who begins his life like other men, who develops, and who finally, by his heroic struggles, achieves divinization. If a Christian projects this sort of future for the world, he will, of course, add that the world's struggle and eventual triumph are possible only through God's power and through faith in Him. Nevertheless this picture is tainted by paganism. It is based on a naïve view about the condition of man and the world which can be appealing to us but which leads to disillusionment in the end. Let us call this picture the "liberal" or Pelagian eschatology.

The most serious error of liberal eschatology is its ignoring of Original Sin. The present world order is one of solidarity in sin and death. There is just no way to attain the state of peace and joy which men desire without radical revision of the whole fabric of the cosmos. The defective structures which have been established in the past ten billion years will continue to guide the course of history until the moment when a power greater than that of angels or men wrests it from its present path and makes a new beginning. To avoid seeing this one must do two things: First, one must minimize the interconnectedness of the cosmos and man's subjection to it; second, one must drastically restrict one's aspirations for happiness. The first step maims the understanding and leaves man a stranger in the cosmos; the second stultifies man's capacity for love and joy and tends to make him a paltry creature with a very low level of aspiration.

A second schematic picture of the future is this: The world and man are incurably sinful and corrupt. Nature and the powers of nature are sinful in the sight of God, and eventually all positive effort and striving will be revealed as useless and empty. In the end man and the world will be utterly frustrated and will die, with either a bang or a whimper. Then when all that is not divine has been revealed as totally insufficient, God will intervene and create from the ruins of the present aeon a new and perfect creation. Let us call this picture the "neo-orthodox" eschatology.

In spite of its emphasis on the greatness of God and the destructive power of sin, neo-orthodox eschatology too is tainted by paganism. This is not to say that it is not intelligent. It is considerably more intelligent than the fatuous hope of liberalism that man can perfect himself. However, like liberalism, it does not correctly understand the pattern of God's dealings with men. That pattern is incarnational, sacramental, and dialogic.

When the fullness of time had come, it was not a human act which brought Jesus Christ into the world but rather the descent of the Holy Spirit upon a simple Jewish girl who, as she herself acknowledged, was unworthy of the honor and powerless to bring it about. Nevertheless this act did not take place without a long preparation which included ten billion years of cosmic evolution and a million years of human striving to rise above the level of the brutes. Neither did it take place without the consent of the maiden who is the virgin daughter of Sion, the final flowering of a millennial tradition which, for all its defects and failures, had in this individual developed the potential created by evolution to a state of marvelous perfection. Countless seeds must die in order that one may germinate; thus the glory and exaltation, the blood and tragedy of Israel's history, flowered at last in a way which was not entirely unworthy of the divine response which it evoked.

Besides the Incarnation the best analogy we have for the Parousia is the Eucharist.[4] In the Mass the bread and wine, fruit of the earth and the vine and the work of human hands, are transformed in response to the faith of the Church into the body and blood of Christ, present in love to nourish and support them. So it will be at the end of time. The whole of creation will be definitively united with Christ and so will find its way to the Father. Without the offering of Cosmic Powers and man, there would be nothing to be transformed. Without the love expressed in this offering, it would be impossible for God's reply to perfect a community of love.

Because it is a history of love between persons, the structure of God's dealings with men is not only incarnational and sacramental but also dialogic. There can be no full personal love without dialogue, without acts on the part of both parties. And in these acts the parties respond to one another, they take into account what has gone before. God's love would not be truly personal and respectful of man if He ignored and did not make use of man's admittedly feeble, but nonetheless real, efforts. Because God is as far exalted above man as heaven above earth there is no comparison, objectively speaking, between the contributions of the two parties. But love tends to make lovers equal, and infinite love can even bridge the gap between God and man.

Liberal eschatology destroys the dialogic pattern of God's action in the world because it makes God's power a subjective possession of man and does not allow God to confront man as a "Thou" who responds to Him from without. Neo-orthodox eschatology also destroys the dialogic pattern because it has God overwhelming rather than responding to man. God's ultimate word of love awaits the full self-expression of the cosmos. When all the potentialities implanted by God in the cosmos have been fully expressed, when all the members of the total community of created persons have spoken their words of response to the gift of self, then God will reply with His final word of judgment, pity, and healing love. The explicit structures which have emerged from the subjective potencies of Cosmic Powers and men will be completed, healed, elevated, and transformed into the new creation, united in the bonds of personal and dialogic love to its creator.

Original Sin has complicated the world and God's plans for men. Human history would have been dialogic even if sin had not occurred. Once God calls man to sonship, i.e., once a supernatural world order has been established, dialogue between man and his Father is the very core of reality. But in a sinless world this dialogue would have proceeded smoothly, without discontinuities or the necessity of doing violence to defective dynamisms in order to make room for perfect ones.

The present situation of man and cosmos is quite different. The defective structures established by sinful actions bring forth bad fruit. It is beyond the power of man and the Cosmic Powers who stand on the side of God to re-establish the lost order of the world. That order needed the cooperation of Satan and those associated with him in order to function properly. Now they not only refuse cooperation but exert a powerful destructive influence. Only the judgment of God can rectify the situation. That judgment will come in due course and it will necessarily have an aspect of violence. The vivid apocalyptic imagery of Scripture is not mistaken. The world cannot be definitively saved unless the evil will of those who oppose the common good of the cosmos is blocked so that it cannot produce its natural effects.

Christians look forward to the day of the Lord with hope and with trepidation. We too are involved in evil and that evil must be purged. To the extent that we relinquish it voluntarily, we have nothing to fear; but to the extent that we try to cling to it, the judgment of God will fall upon us also as violence.

5. THE WORKS OF MAN

As we have seen, Christian hope cannot be simply an attitude of expectation of divine action. We ourselves have a role to play. But why are our human strivings important? It is difficult to say precisely why. We have a heuristic anticipation based on faith, hope, and the past pattern of God's dealings with men. We know that we have entered into the hopes and plans of the Cosmic Powers who were originally entrusted by God with the task of building the cosmos and of the Son of Man who came to save us when those plans went awry. With Teilhard de Chardin we can believe that the divine goodness and power are great enough to give to creatures the gift of being real creators of something of permanent value, even though we cannot do it unaided. Nevertheless we do not know the shape of the future or the way in which God will use it. As far as the permanent worth of the works of man is concerned we can only live in hope.

But the theory of symbolic activity does enable us to say something definite about the subjective side of human activity. Our human strivings are important at least because of this: The decision which is love can come into being only in the emanation of a symbol from the subjective potency of the person. That symbol is the effort, the total behavior which aims at creating here in the cosmos a community of love in which men and all intelligent creatures are united under the fatherhood of God. Man has no field other than the cosmos within which to symbolize. If we do not strive to organize matter and human society, to fulfill to the uttermost the naturally good potentialities of our own natures and those of our fellows, we cannot act or love at all. Furthermore, only by external action can we communicate our love to others. The love of Christ can be made real to men only in the loving action of other men for their welfare, and the human community can be welded into unity only in the cooperative effort to achieve the common good of all.

Thus even if the hope that the external works of man will contribute in some way to the new creation is mistaken, it is certain that the subjective actualization which arises in the effort to create them is essential. The heavenly Jerusalem exists now only in the intention of God and will descend to earth only at the Parousia. But we can and must struggle to build a just earthly city, a realistic Utopia which takes into account the actual situation and the resources available. Such a Utopia needs only good will for its realization.

"Only good will"—such a phrase may well bring a bitter smile to the face of a twentieth-century man. Good will is precisely what we lack. There is no dearth of intelligent people with good ideas; the real problem is our unwillingness to subordinate our own narrowly personal interests to the common good. The Christian doctrine of Original Sin teaches us that moral evil is not simply ignorance; no amount of knowledge can insure good moral choices. In fact, as we have seen, good moral choice is ultimately identical with the creation of true knowledge. The doctrine of Original Sin also teaches us that man is so involved with sin from the first moment of his life that he is bound to make it his own to some extent. The probability of his avoiding it completely has measure zero,[5] even though, since man is free, such avoidance is not simply impossible.

Therefore even a realistic Utopia is an ideal rather than an attainable goal. It is something which we can approximate to the extent that we are willing to love the common good. Doubtless we will always fall short of the good which could be obtained even in the present world order, but we can hope that somewhere, in a limited arena, for a brief period of time, we can achieve a recognizable approximation to it. Such a hope is what God commands us to have. We begin where we are, with

the resources we have, and use them to the full extent of our
abilities. Until the return of our Lord, we work with the ten,
or five, or one talent which He has entrusted to us. In so
doing we necessarily achieve something, even if our external
projects are completely frustrated, for we bring into being
within ourselves the personal attitudes which speak to the
heart of God and evoke in Him the response of salvific judgment for ourselves and others.

These personal attitudes are, in the first instance, attitudes of
love, and this we should remember when we speak of trying to
build an earthly Utopia or a just social order. Utopia can be a
happy home for human beings only if it is an order based on
love. The common good is the good of oneself and of others,
and one wills it because one loves oneself and others. Social
systems which achieve it are ultimately complex ways of doing
good to individual persons.

Though love is the very heart of Utopia, we are speaking of
an earthly commonwealth, that is to say, of a social system which
is part of the present sinful world order. Therefore like all the
systems which now exist, it will have to depend very heavily
on mechanisms. The organismic regulation of the total system
is the mutual love of its citizens. But this love cannot do
everything by itself. To be effective, it must organize and
control social mechanisms which run on lower motives, such
as economic, political, and other interests. Furthermore, the
planners of Utopia must take into account human sinfulness,
competitiveness, pride, sloth, etc. Fear and coercion are inevitable components of any practical Utopia. Without them
too much strain is exerted on altruistic love, and things begin
to collapse. There is a strong tendency among Christian builders
of Utopia to confuse the earthly city with the heavenly. In the
Jerusalem above, which is our true home, economic pressure
is unknown, and the judge, policeman, and soldier have no
place. But here below they are indispensable. To refuse their
services is to refuse the task of constructing a workable social
system. Mechanisms which are not controlled by love lead to
an inhuman, machinelike society, but love without mechanisms
results in chaos and the extinction of love.

The earthly hopes of men are ordered to and informed by

their eschatological hopes. Without the latter the former can never be stable or enduring, for the chance of their being realized as fully as we wish is slight, and even if they were realized, such realization would not be totally satisfying. Many callow youths are capable of beginning idealistic Utopian projects. But as the years go by and painful and disillusioning experience accumulates, men are tempted to abandon youthful dreams or to transform them into means of self-aggrandizement. The natural hopes of most men must either perish or be assumed into something greater. The Christian who subsumes them under his eschatological hope can endure their apparent frustration because he knows that no effort he makes is really in vain.

6. THE HOPE WHICH IS OURSELVES

Our hope is a heuristic anticipation, a knowing desire or a desireful implicit knowledge of what is to come. In Polanyi's words, "it resembles not the dwelling within a great theory of which we enjoy the complete understanding, nor an immersion in the pattern of a musical masterpiece, but the heuristic upsurge which strives to break through the accepted frameworks of thought, guided by the intimations of discoveries still beyond our horizon. Christian worship sustains, as it were, an eternal never to be consummated hunch; a heuristic vision which is accepted for the sake of its unresolvable tension. It is like an obsession with a problem known to be insoluble, which yet follows against reason, unswervingly, the heuristic command: 'Look at the unknown!' " [6]

These words are a magnificent expression of the nature of hope, but a hope which resembles Job's more than Paul's or John's. On the one hand they catch something of the passionate yearning for the light which echoes in the Dominican motto of St. Thomas Aquinas, *Veritas,* as well as in the lives and writings of men like Kepler and Einstein. On the other hand they lack the assurance which comes to a Christian who lives in the bosom of the Church. That assurance does not weaken the heuristic tension of hope but brings confidence that it will in fact be satisfied. Christian hope contains a component of joyous and confident fulfillment because the light has already dawned within and the sweet anointing of the Spirit soothes the tension of our powers as they strain toward what is yet to come, enabling us to give ourselves more fully to the otherwise unbearable dynamism.

In the final analysis, what is man? What am I? I am an embodied hope, an embodied love which yearns for its own fulfillment. "The living flame of love" [7]—that which the mystics have experienced to the full, that are we all even when we know it not. The human spirit is a grandeur far greater than the spreading galaxies which make our whole planet and all our material creations a mere speck of dust in the immensity of the cosmos. Hope and love are stronger than death; many waters, even the waters of death and primeval chaos, cannot extinguish them (Canticle 8:7). This is true in every man but above all in *The Man*, Jesus the Christ. In Him the truth about man has been revealed. And this happened most fully in the moment when He went down into death crying with a loud voice, "Father, into your hands I commit my spirit" (Lk 23:46). In His footsteps we shall all follow sooner or later.

Christ is also the full revelation of the Father, a revelation most powerfully expressed perhaps in the statement "God is love" (1 Jn 4:8). "Only love can be believed." [8] Ultimately what we believe in is the love of God appearing to us in Christ Jesus. This love is the answer to the heuristic anticipation which is myself. It is what I have known in anticipation from my first moment and have been searching for ever since. The gospel of its appearance (as witnessed to in the Church) is therefore self-validating. In the moment when a man is (effectively) confronted with the gospel and succeeds in penetrating into his own center and taking as the core of his instantaneous noetic structure his own essential self, in that moment he knows that the Christian message was not invented by men but comes from the wellspring of his own being. The "two hands of God," [9] one within and one without, close round him in the beginnings of the embrace for which he was made.

So expected and yet so unexpected! "Lord, who has believed our report?" (Jn 12:38, Is 53:1). What we have always known and yet so undreamed of in its concrete reality. My decision about God and about Christ is a decision about myself. Am I absurd, an obscene contradiction who defaces the earth for a while and then putrifies? Or is the freshness of my childhood yearning, a yearning which persists within me in spite of all that I or the world can do to excise it, the truth about the world? In the moment when I decide, I know about God and Christ, and in the moment when I know about God and Christ, I decide. If I am not absurd, then it is true that God has appeared to me in Christ; if God has appeared to me in Christ, I am not absurd.

When our inmost self has been explicitated in an act of faith, our noetic structure is transformed and our life begins to develop within a new horizon. In hope we live a new life, that is, we understand and desire in a new way and so move toward a goal unknown to those without hope. Through hope we know that we must work for something which we cannot realize through our own efforts alone. We know also that our work will not be in vain because it will be taken up and completed by God. Thus the Christian can have an attitude which combines humility and confidence and so avoids both delusion and despair.

The risen Christ is the visible sign of the reality of our hopes. The cosmos still awaits the ultimate transforming word of love from the Father, but the resurrection of Christ is the incarnate promise of its coming. God's creative decision was from the very beginning the sphere of real possibility and disposable power within which the cosmos could come to be. But all that was latent there was not explicitated in the natures of creatures at the start. The potentialities present within the Cosmic Powers were not equal to the full scope of God's intentions. Only with the life, death, and resurrection of Christ has the fullness of the plan hidden in God from the beginning been made known. With the resurrection, the creative will of God has assumed a new dimension of graciousness toward us and brought into being a new term, a new grace, within each person. This grace is perhaps best described as a new hope, a new orientation toward the future and the final union of creation with God. The core of the universe is the society of persons within it, and the very essence of a person is his subjective heuristic anticipations and desires. Therefore in transforming the fundamental subjective hopes of men and angels and joining them to Christ's own, the resurrection has transformed the basic ontological structures of reality. The formal laws of nature and the mechanisms which operate according to them may be the same, but the organismic regulation which the society of persons, angelic and human, exerts is different, and so the creative aspect of nature's operations is different.

Teilhard's vision of the world as Eucharist was no illusion.[10] Whitehead defined the human body as "that region of the world which is the primary field of human expression." [11] With the resurrection the persons in the cosmos, at least those of good will, have become the Whole Christ by their common participation in the Spirit. Their ultimate field of expression is the entire universe which is thus their body in a very real, though analogous, sense. During the present age of the world the organismic regulation of the Whole Christ is directed to the salvation of men by the way of the cross rather than by immediate glorification. The ill will of those who reject the good and the discordant structures which this will has established in the past, and continues to establish today, remain effective. Nevertheless this is a redeemed world in which all things work together for the good of those who love God (Rom 8:28). By patience we shall possess our souls and arrive at last at the moment when all things will be made new and God will wipe every tear from our eyes (Rev 21:4).

NOTES

1. MacGowan, p. 369.
2. The idea of a state of fulfillment in which further development is excluded is incomprehensible and distasteful to many moderns, partly because we have never experienced such a state. In the present age not to develop is to deteriorate. I think it may also result from psychological pathology since the idea of just *being* without struggle threatens neurotic defenses. See Janov, Chapter 11.
3. It is also proper to God insofar as He has involved Himself with us. Nevertheless His essential nature remains eternal.
4. See Teilhard (III).
5. "Measure" is a term used in the theory of probability. It is the generalization of the idea of area. Consider a square with unit side. It has measure (and area) 1. The lower right hand quarter has measure

(and area) ¼. Now consider a single one of the uncountably infinite number of points which compose the square. It has measure 0; in fact, so does any countable set of points. It takes an uncountable number of points to have nonzero measure. If one chooses a point from the square at random, the probability of picking any particular point is zero (probability being in direct proportion to measure); nevertheless one is certain to pick some point, so zero probability and impossibility are not the same.

6. POLANYI (I), p. 199.
7. JOHN OF THE CROSS, p. 717. My use of the phrase is not precisely that of the poem, though I believe that there is a relationship.
8. BALTHASAR, p. 83.
9. TEILHARD (II), p. 78.
10. TEILHARD (III), p. 46. Note that the understanding of symbolizing activity which we have developed can be used to clarify the intelligibility latent in the terms "transubstantiation" and "transignification."
11. WHITEHEAD (III), p. 30.

Bibliography

Adler, Mortimer. *The Difference of Man and the Difference It Makes.* Cleveland: World, 1968.

Aquinas, St. Thomas. *Summa Theologica.*

Ardrey, Robert. (I) *African Genesis.* New York: Atheneum, 1961.

———. (II) *The Territorial Imperative.* New York: Atheneum, 1967.

Balthasar, Hans Urs von. *Love Alone.* New York: Herder, 1969.

Barbour, Ian. *Issues in Science and Religion.* Englewood Cliffs: Prentice-Hall, 1966.

Berger, Peter. (I) *The Sacred Canopy.* New York: Doubleday Anchor, 1969.

———. (II) *A Rumor of Angels.* New York: Doubleday Anchor, 1969.

Bertalanffy, Ludwig von. *General System Theory.* New York: Braziller, 1968.

Bilaniuk, Olexa-Myron, and E. C. George Sudarshan. "Particles Beyond the Light Barrier." *Physics Today* 22, No. 5 (1969), 43–51.

Brans, C. H., and R. H. Dicke. "Mach's Principle and a Relativistic Theory of Gravitation." *Physical Review,* Second Series, No. 3 (1 November 1961), 925–935; (Brans, C. H.) "Mach's Principle and a Relativistic Theory of Gravitation. II." *Physical Review,* Second Series, No. 6 (15 March 1962), 2194–2201.

Buckley, Walter (ed.). *Modern Systems Research for the Behavioral Scientist.* Chicago: Aldine, 1968.

Camus, Albert. (I) *The Plague.* New York: Knopf, 1960.

———. (II) *The Rebel.* New York: Knopf, 1954.

Chew, Geoffrey. (I) "The Dubious Role of the Space-Time Continuum in Microscopic Physics." *Science Progress* 51, 529.

———. (II) "Hadron Bootstrap: Triumph or Frustration?" *Physics Today* 23, No. 10 (October 1970), 23–28.

Coreth, Emerich. *Metaphysics.* New York: Herder, 1968.

Décarie, Thérèse. *Intelligence et affectivité chez le jeune enfant*. Neuchâtel: Delachaux et Niestlé, 1962.

DeWitt, Bryce. (I) "Quantum Mechanics and Reality." *Physics Today* 23, No. 9 (September 1970), 30–35.

DeWitt, Bryce, et al. (II) "Quantum Mechanics Debate." *Physics Today* 24, No. 4 (April 1971), 36–44.

Dobzhansky, Theodosius. *Mankind Evolving*. New Haven: Yale University Press, 1962.

Eden, Murray. "Inadequacies of Neo-Darwinian Evolution as a Scientific Theory." *Mathematical Challenges to the Neo-Darwinian Interpretation of Evolution*, edd. Paul S. Moorhead and Martin M. Kaplan. Philadelphia: Wistar Institute Press, 1967.

Eliade, Mircea. *The Sacred and the Profane*. New York: Harcourt, Brace and World, 1959.

Feynman, Richard. *The Character of Physical Law*. Cambridge: M.I.T. Press, 1965.

Fitzmyer, Joseph. "Pauline Theology." *The Jerome Biblical Commentary*, edd. Raymond Brown, Joseph Fitzmyer, and Roland Murphy. Englewood Cliffs: Prentice-Hall, 1968; pp. 800ff., esp. p. 824, No. 140.

Flavell, John. *The Developmental Psychology of Jean Piaget*. New York: Van Nostrand, 1963.

Furth, Hans. *Piaget and Knowledge*. Englewood Cliffs: Prentice-Hall, 1969.

Gendlin, Eugene T. "A Small, Still Voice." *Psychology Today* 4, No. 1 (June 1970), 56–59.

Gilson, Étienne. *The Christian Philosophy of St. Thomas Aquinas*. New York: Random House, 1956.

Grelot, Pierre. "Réflexions sur le problème du péché original." *Nouvelle Revue Théologique* 89 (1967), 337–375, 449–484.

Grene, Marjorie. *The Knower and the Known*. New York: Basic Books, 1966.

Guardini, Romano. *The Lord*. Chicago: Regnery, 1954.

Harary, Frank, Robert Norman, and Dorwin Cartwright. *Structural Models*. New York: Wiley, 1965.

Harris, Errol. *The Foundations of Metaphysics in Science*. New York: Humanities Press, 1965.

Harrison, Edward. "The Early Universe." *Physics Today* 21, No. 6 (June 1968), 31–39.

Hebb, Donald. *The Organization of Behavior: A Neurophysiological Theory*. New York: Wiley, 1949.

Heelan, Patrick. *Quantum Mechanics and Objectivity*. The Hague: Nijhoff, 1965.

Heisenberg, Werner. (I) *Physics and Philosophy*. New York: Harper, 1958.
——. (II) *Physics and Beyond*. New York: Harper and Row, 1971.
Hilbert, David. "On the Infinite." *Philosophy and Mathematics*, edd. Paul Benacerraf and Hilary Putnam. Englewood Cliffs: Prentice-Hall, 1964.
Jaki, Stanley. (I) *The Relevance of Physics*. Chicago: University of Chicago Press, 1966.
——. (II) *Brain, Mind and Computers*. New York: Herder, 1969.
Janov, Arthur. *The Primal Scream*. New York: Delta, 1971.
John of the Cross, St. *The Collected Works of St. John of the Cross*. Trans. Kieran Kavanaugh and Otilio Rodriquez. Garden City: Doubleday, 1964.
Kellogg, Winthrop, and Luella Kellogg. *The Ape and the Child*. New York: McGraw-Hill, 1933.
Koestler, Arthur. *The Act of Creation*. New York: Macmillan, 1964.
Kuhn, Thomas. *The Structure of Scientific Revolutions*. Chicago: University of Chicago Press, 1970.
Lakatos, Imre, and Alan Musgrave (edd.). *Criticism and the Growth of Knowledge*. New York: Cambridge University Press, 1970.
Lonergan, Bernard. *Insight*. New York: Philosophical Library, 1957.
Lorenz, Konrad. (I) *Physiological Mechanisms in Animal Behavior*. Symposia of the Society for Experimental Biology, No. 4. New York: Cambridge University Press, 1950.
——. (II) *On Aggression*. New York: Bantam, 1967.
MacGowan, Roger, and Frederick Ordway. *Intelligence in the Universe*. Englewood Cliffs: Prentice-Hall, 1966.
Magee, John. *Religion and Modern Man*. New York: Harper and Row, 1967.
Maritain, Jacques. *Approaches to God*. New York: Harper, 1954.
McKenzie, John. *Dictionary of the Bible*. Milwaukee: Bruce, 1965.
Misner, Charles. "Mass as a Form of Vacuum." *The Concept of Matter*, ed. Ernan McMullin. Notre Dame: University of Notre Dame Press, 1965.
Montagu, Ashley (ed.). *Man and Aggression*. New York: Oxford University Press, 1968.
Moorhead, Paul S., and Martin M. Kaplan (edd.). *Mathematical Challenges to the Neo-Darwinian Interpretation of Evolution*. Philadelphia: Wistar Institute Press, 1967.

Peebles, P. J. E., and David T. Wilkinson. "The Primeval Fireball." *Scientific American* 216, No. 6 (June 1967), 28–37.
Pendergast, Richard. (I) "The Supernatural Existential, Human Generation and Original Sin." *Downside Review* 82 (1964), 1–24.
———. (II) "Terrestrial and Cosmic Polygenism." *Downside Review* 82 (1964), 189–198.
Perls, Frederick, et al. *Gestalt Therapy.* New York: Delta, 1965.
Piaget, Jean, and Bärbel Inhelder. (I) *The Psychology of the Child.* New York: Basic Books, 1969.
Piaget, Jean. (II) *Genetic Epistemology.* New York: Columbia University Press, 1970.
Poincaré, Henri. "Mathematical Creation." *Mathematics in the Modern World.* San Francisco: Freeman, 1968.
Polanyi, Michael. (I) *Personal Knowledge.* Chicago: University of Chicago Press, 1958.
———. (II) *The Study of Man.* Chicago: University of Chicago Press, 1959.
———. (III) *The Tacit Dimension.* Garden City: Doubleday Anchor, 1967.
Rahner, Karl. (I) *On the Theology of Death.* New York: Herder, 1961.
———. (II) *The Eternal Year.* Baltimore: Helicon, 1965.
———. (III) "Theology of the Symbol." *Theological Investigations* IV. Baltimore: Helicon, 1966.
———. (IV) "Christology Within an Evolutionary View." *Theological Investigations* V. Baltimore: Helicon, 1966.
Renckens, Henricus. *Israel's Concept of the Beginning.* New York: Herder, 1964.
Schlier, Heinrich. *Principalities and Powers in the New Testament.* New York: Herder, 1961.
Schoonenberg, Piet. *Man and Sin.* Notre Dame: University of Notre Dame Press, 1965.
Schützenberger, Marcel. "Algorithms and the Neo-Darwinian Theory of Evolution." *Mathematical Challenges to the Neo-Darwinian Interpretation of Evolution,* edd. Paul S. Moorhead and Martin M. Kaplan. Philadelphia: Wistar Institute Press, 1967.
Sciama, Dennis. *The Unity of the Universe.* Garden City: Doubleday, 1959.
Shannon, Claude, and Warren Weaver. *The Mathematical Theory of Communication.* Urbana: University of Illinois Press, 1949.
Storr, Anthony. *Human Aggression.* New York: Atheneum, 1968.

Teilhard de Chardin, Pierre. (I) *The Phenomenon of Man.* New York: Harper Torchbook, 1965.
——. (II) *The Divine Milieu.* New York: Harper Torchbook, 1965.
——. (III) *Hymn of the Universe.* New York: Harper Colophon Book, 1969.
Tinbergen, Nikolaas. *The Study of Instinct.* New York: Oxford University Press, 1951.
Weisskopf, Victor. "Three Steps in the Structure of Matter." *Physics Today* 23, No. 8 (August 1970), 17–24.
Wheeler, John. *Geometrodynamics.* New York: Academic Press, 1962.
Whitehead, Alfred North. (I) *Process and Reality.* New York: Macmillan, 1929.
——. (II) *Science and the Modern World.* New York: Macmillan, 1925.
——. (III) *Modes of Thought.* New York: Macmillan, 1938.